*Industrial Energy Management Strategies:
Creating a Culture of Continuous Improvement*

Industrial Energy Management Strategies: Creating a Culture of Continuous Improvement

By Kaushik Bhattacharjee

THE FAIRMONT PRESS, INC.

CRC Press
Taylor & Francis Group

Library of Congress Cataloging-in-Publication Data

Names: Bhattacharjee, Kaushik, 1970- author.
Title: Industrial energy management strategies : creating a culture of
  continuous improvement / by Kaushik Bhattacharjee.
Description: Lilburn, GA : Fairmont Press, Inc., [2017] | Includes
  bibliographical references and index.
Identifiers: LCCN 2017039964| ISBN 0881737690 (alk. paper) | ISBN 0881737704
  (electronic) | ISBN 9780815380016 (taylor & francis distribution : alk.
  paper)
Subjects: LCSH: Energy conservation. | Industrial efficiency.
Classification: LCC TJ163.3 .B49 2017 | DDC 658.2/6--dc23 LC record available
  at https://lccn.loc.gov/2017039964

*Industrial Energy Management Strategies: Creating a Culture of Continuous
  Improvement / Kaushik Bhattacharjee*
©2018 by The Fairmont Press, Inc. All rights reserved. No part of this publication
may be reproduced or transmitted in any form or by any means, electronic or
mechanical, including photocopy, recording, or any information storage and
retrieval system, without permission in writing from the publisher.

Published by The Fairmont Press, Inc.
700 Indian Trail
Lilburn, GA 30047
tel: 770-925-9388; fax: 770-381-9865
http://www.fairmontpress.com

Distributed by Taylor & Francis Group LLC
6000 Broken Sound Parkway NW, Suite 300
Boca Raton, FL 33487, USA
E-mail: orders@crcpress.com

Distributed by Taylor & Francis Group LLC
23-25 Blades Court
Deodar Road
London SW15 2NU, UK
E-mail: uk.tandf@thomsonpublishingservices.co.uk

Printed in the United States of America
10 9 8 7 6 5 4 3 2 1

ISBN: 0881737690 (The Fairmont Press, Inc.)
ISBN: 9780815380016 (Taylor & Francis Group LLC)

While every effort is made to provide dependable information, the publisher, authors, and
editors cannot be held responsible for any errors or omissions.

The views expressed herein do not necessarily reflect those of the publisher.

Table of Contents

Chapter 1
Energy Analysis 1
    Introduction 1
    Utility Bill Analysis............................. 2

Chapter 2
Energy Conservation Opportunities 23
    Energy Conservation Opportunities in
        Industrial Ventilation Systems................ 24
    Heat Recovery Opportunities in
        Manufacturing Systems 36
    Process Heat Recovery System 40
    Energy Optimization in Chilled Water Systems........ 44
    Cost-Reduction Strategy 45
    Energy Optimization in
        an Industrial Cooling Water System 50
    Energy Optimization in Industrial Refrigeration System . . . 55

Chapter 3
Lean Manufacturing Principles and
Their Impact on Process Energy Efficiency 77
    Introduction 77
    Lean Manufacturing Basics 78
    More on Andon................................. 93
    Andon Equipment and Vendors 104
    Summary and Conclusions 128
    Glossary 130

Chapter 4
Operational Savings................................. 139
    Energy Traffic Light Program 140
    Common Operational Measures................... 141
    Verification of Operational Savings................ 151

Chapter 5
Selling Energy Projects **155**
    Using the Correct Matrix. 157
    Case Studies Illustrate Opportunities
        for Multiple Utility Savings 162

Chapter 6
Measurement and Verification of Industrial
Energy Conservation Projects **169**
    Introduction 169
    Case Studies 169

Chapter 7
Project Pitfalls **191**
    Energy Management and Energy Efficiency Projects. 191
    Project Pitfalls 191
    Post-project Operating Strategy 197
    Implementation Sequence on An Optimization Project . . . 198
    M&V 200

Chapter 8
Energy Management Best Practices **205**
    The Energy Management Continuum 205
    Necessary Elements for the Success of
        a Sustainable Energy Program 206
    Some of the Key Initiatives to Address Barriers 210
    Energy Management Review Contours 224
    Improving or Implementing
        an Energy Management Program. 228

Chapter 9
Application of Energy Monitoring and Targeting for
Industrial Plants **229**
    Components of an Energy Management
    Information System. 229
    CUSUM Analysis 230
    RETScreen Expert. 233
    Monitoring and Targeting 264
    Case Studies on Developing an M&T System
        for Small and Mid-sized Facilities 267

Chapter 10
Energy Management Innovation
Time-shared Energy Manager 275
    Introduction 275
    Barriers to Implementing An
        Energy Management Program in SME. 275
    The role of the energy manager. 276
    Case Study of the Time-shared Energy Manager 278

Index **289**

Foreword

The importance of energy management has grown in recent years due to the heightened awareness of the impact of energy use on the environment, and its very real impact on a company's bottom line. With the need for energy management gaining wider acceptance, the need for a book that can provide a detailed and knowledgeable reference for those engaged in the field, or just starting out, is long overdue. *Industrial Energy Management Strategies: Creating a Culture of Continuous Improvement*:

1. Provides a practical approach to implementing energy management programs using case studies and real-world experience.

2. Introduces and clearly explains new areas of development that are gaining importance in the industry, such as CUSUM and multivariate regression analysis.

3. Provides coverage of all systems and standards that affect energy management, including ISO50001, EMIS, Industrial Refrigeration, Cooling Water System and Industrial Ventilation System.

4. Addresses all aspects of achieving effective energy management including technical, organizational, and behavioural considerations.

5. Is designed as a quick reference guide for practising energy managers, as well as providing the necessary background and skills for operators, managers, and students new to the field.

Following are some of the highlights of the book.

Chapter 1—Energy Analysis
The chapter shows how the components of energy analysis, such as utility bills, energy use breakdown, study of process maps, utility interval data analysis, etc., are analyzed together to understand energy use and issues related to energy management.

Chapter 2— Energy Conservation Opportunities in Common Process Auxiliaries

A very basic conceptual description would be carried out for the following energy-consuming system(s). Case studies of common opportunities are presented for the following areas. Each case study has detailed explanations of calculation and assumption used; the calculation can be replicated for similar situations by the reader:

- Cooling water system
- Ventilation system
- Compressed air system
- Lighting system
- Waste heat recovery
- Industrial refrigeration system
- Vacuum system
- Steam system

Chapter 4—Operational Savings

There can be many operational energy reduction opportunities in different types of manufacturing facilities. This chapter presents some commonly used energy reduction opportunities with case studies and calculation of energy savings (where applicable).

Chapter 5—Selling Energy Projects

Successful strategies for selling energy projects to senior management using case studies that include all the relevant benefits (including non-energy benefits) of the project are shown.

Chapter 6—Measurement and Verification (M&V)

This chapter discusses measurement and verification carried out for some of the implemented electricity conservation and demand-side management projects. In many of these cases, Option A, Option B, and Option C from the IPMVP (International Performance Measurement and Verification Protocol) guidelines are followed. Elements of the M&V plans are discussed. Some of the issues, like metering and methods used for determining variations of electricity for different operating conditions are highlighted. The benefits of the M&V process to all the stakeholders are discussed.

Chapter 7—Project Pitfalls
Case studies illustrate some common project pitfalls while implementing energy projects, including poor analysis of all available energy conservation options, weak baseline measurements, failure to define project scope, and failure to maximize utility incentives for the project.

Chapter 8—Energy Management Best Practices
The chapter shares best practices supporting a continuous energy performance improvement.

Chapter 9—Monitoring and Targeting (M&T)
Introduction of concepts related to monitoring and targeting, such as CUSUM analysis. Case study examples will be included to demonstrate uses of CUSUM analysis in monitoring energy consumption and target setting.

Chapter 10— Energy Management Innovation:
The Time Shared Energy Manager Model
Small- and mid-sized manufacturing industries lack the resources to maintain a full-time energy manager, so energy management functions and responsibilities are left mostly to the plant manager. The plant manager, already loaded with numerous responsibilities, can hardly devote time and attention to the energy management function. In many cases, they also lack the skills to carry out such a function. At best, a company hires an energy service firm to identify and implement one or two energy reduction measures. This approach does not lead to sustainable energy reduction for the plant. A qualified and experienced energy manager can offer energy management services to a number of small- and mid-sized industries on an ongoing basis. The service can be affordable to small- and mid-sized energy consumers, as the total service cost is shared between different companies. Implementation experience of this model from the Ontario SaveONenergy program is presented.

I would like to thank:

Dr. Wayne Turner for reviewing the proposal of this book and providing insightful comments in the initial stages.

Steve Doty, a very good friend and author of *Commercial Energy audit reference handbook*, for his support and encouragement to help complete this endeavor.

Michel Stocky, for providing input on energy management best practices and authoring the chapter on lean manufacturing.

Jon Feldman from IESO has encouraged me to share my experiences as a Roving Energy Manager, which help me develop insights that have been shared in this book.

Steven Dixon from Knowenergy who has helped me understand and apply concepts of monitoring and targeting.

Justin Macdonald for reviewing and providing his insight on refrigeration and chilled water system.

Brian Fountain for providing valuable insight on chilled water system analysis.

Trishul and Tanu Dave for taking the time to help with the graphics in the book.

I would like to thank my brother Kabir Bhattacharjya for reviewing and providing very valuable inputs while writing the book, and I would also like to thank my parents Dinesh Bhattacharjya and Binita Bhattacharjya for setting me in the path.

Last but not the least my wife, Kuntala, and Anubhuti for their uncomplaining support to complete the book.

Chapter 1
Energy Analysis

INTRODUCTION

Energy analysis is important to understanding how energy is used in a facility. It can provide information on the cost of operating various equipment, information on the peak and base load of the facility, and other related matters. Such an analysis would provide information on the major issues affecting energy consumption in a facility.

Components of energy analysis include utility bill analysis, energy use breakdown, study of process maps, energy measurements, and utility interval data analysis. Information for all these components can be analyzed together to understand energy use and issues related to energy management. This information can also help in finding opportunities for energy reduction.

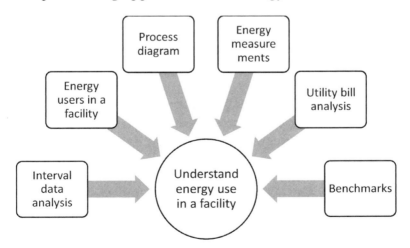

Figure 1-1. Energy Analysis

1

UTILITY BILL ANALYSIS

Utility bills provide information on energy consumption and spending for manufacturing plants. They also provide information on peak demand for a plant, along with unit energy price and other factors, such as distribution charges. In a deregulated market, this information can help in negotiating better prices with energy suppliers. There also might be other opportunities where a manufacturing plan can reduce costs, such as power factor penalty. The utility bills also provide information on percentage cost breakdown for various fuel sources.

Energy Benchmarks

Energy benchmarks are available for some manufacturing sectors, while nature and variation of manufacturing processes make it difficult to obtain these for others. RETSCREEN Expert a clean energy management tool developed by Natural Resource Canada, provides these values for some sectors. The energy consumption for a facility can be compared with benchmarks to determine potential energy savings.

Energy Use Patterns

Annual energy use patterns can be determined from utility bills, and these patterns can identify abnormalities in consumption patterns and highlight the seasonal (nonseasonal) nature of consumption. Figure 1-2 shows an energy consumption pattern for a manufacturing facility.

There are several variables that drive energy consumption in a manufacturing plant, process or system. Examples of energy drivers are production, CDD (cooling degree days), dry bulb temperature, etc. Sometimes more than one driver is responsible for energy consumption in a facility. Figure 1-3 plots energy consumption and production in a manufacturing facility. As clearly seen, the energy consumption follows a patterns close to production. The relationship between independent and dependent variables can be established based on regression models. In some cases poor correlation between variables can point toward

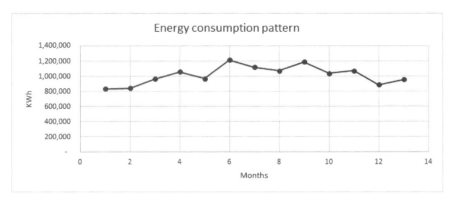

Figure 1-2. Energy Consumption Patterns

improper controls, etc. Energy drivers are useful in developing energy performance models.

Energy Measurements

Several instruments can measure energy and related parameters. These instruments can help determine consumption of equipment, identify areas of energy waste, and reveal optimization opportunities. Examples of some of the measuring instruments are given below.

Amp Measurements

An ammeter is a clamp-on device for checking current draw through one cable, and it can also measure voltage. These are generally good for calculating power draw on small motors which do not show variation in the operating load. They can also measure amp draw on lighting circuits which can be used to calculate lighting power.

Power Analyzers

These meters can measure power and energy for various types of electrical equipment. Power analyzers generally come with logging capabilities. In general, most power meters measure both instantaneous power and have the capability of logging. However, the meters need to be connected correctly, which is one of the most common sources for errors stemming from incorrect

Industrial Energy Management Strategies

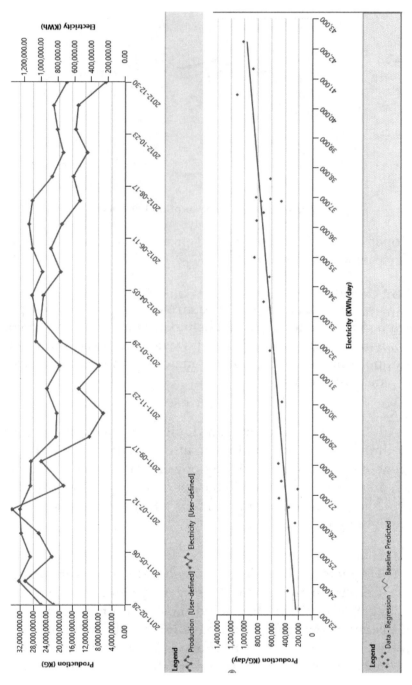

Figure 1-3. Energy Consumption Drivers

Energy Analysis

Figure 1-4. Ammeter

current transformer direction. Invariably, most power analyzers require the current transformers (CTs) to be placed in such a way that the direction of the current is positive or negative. General convention dictates that all three CTs be negative or positive (for three-phase power measurements). In many power analyzers, the total power will not add up if the CTs are connected incorrectly.

For motor control centers, care should be taken to measure the total power. In many cases, more than one cable is connected to the bus, and a common source of error is that the CTs are placed on only one of the cables, which record the total power as being less. In general, it is always advisable to determine the motor's horsepower before measurements and perform a check that the measurement is correct.

Air Flow Meters

There are two types of air flow meters. First is the wheel-type anemometer with a round fan that spins as air flows through it; its display reads the air velocity. Second is a hotwire that measures air velocity. The instrument can be used to check various locations across a vent or duct, then average the velocity and multiply by the cross-sectional area to get an estimate of the CFM air flow.

Figure 1-5. Power Analyzers

Portable Water Flow Meters

Portable water flow meters measure water flow using probes attached to the pipes. The flow meters measure the velocity of water based on ultrasonic principles and will calculate flow based on pipe size, diameter, and other information. For example, portable flow meters are useful in measuring chilled water tons and studying pump system flow. The flow meter can also help in determining pumps' operating efficiency.

Combustion Analyzers

Combustion analyzers can be used to measure furnace efficiency, using flue gas oxygen content, flue gas temperature, ambient temperature, and flue gas carbon monoxide. Based on these measurements, the combustion analyzers can calculate combustion efficiency. For optimization, combustion analyzers can determine flue gas carbon dioxide content and excess air content.

Energy Analysis

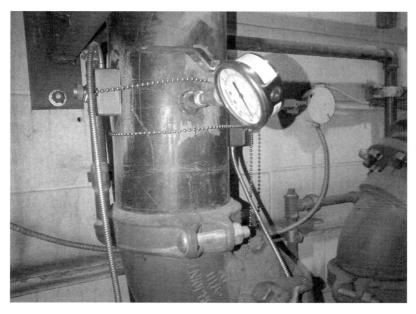

Figure 1-6. Flow Meter (sensors)

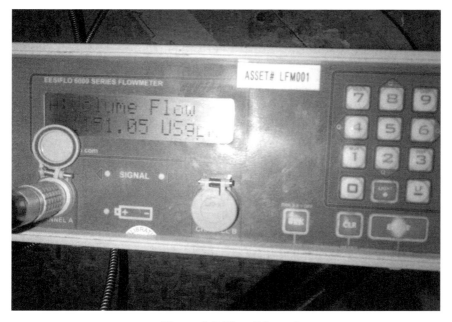

Figure 1-7. Flow Meter, 2

Ultrasonic Leak Detection

Ultrasonic leak detectors are used to detect compressed air leaks. The leak detector essentially amplifies the sound of leakages that would not be otherwise detected because of surrounding noise. The general method to determine leaks is to find gross leaks and then adjust the sensitivity to pinpoint the specific leaks in the air distribution system.

Light Meter

A light meter has an electric eye on its spiral cord placed horizontally on its face to measure the foot-candle of illumination. You can spot check lighting levels with this device.

Low Cost Energy Loggers

Use of low cost energy loggers can reveal much information on equipment idling and can also be used to develop energy consumption profiles for equipment in a facility.

Metering Issues

Below are some metering issues that need to be addressed during meter installation for power measurement.

Measuring Power at Large Feeders

For motor loads, power is generally balanced, so single phase monitoring can be done to observe load profile, etc. The selection of CTs for measurements is important. In general, some of the CTs can measure accurately within 10% of the rated amps of the CT. For example, while measuring input current of a 20HP three-phase induction motor, it is advisable to select a 50A CT as opposed to 100A CT. The rated amps for a 20HP, 575 V induction motor is 22A. A 50A CT can measure amps accurately between 50 and 5 (10%) A, while a 100 A CT would measure amps accurately between 100 to 10A (10%), so with a 50A CT, it would accurately measure amps from 5A while with the 100A CT, it would measure accurately from 10A.

Energy Analysis

Figure 1-8. Amps logger

Figure 1-9. Occupancy Loggers

Data logging on the main power feed can always be accompanied by spot measurements of some of the individual pieces of equipment. This can help during the analysis of trend logs. Figure 1-10 shows logged amperage from compressor measurements in a rooftop unit; the amperage of the two-stage compressors is seen from the figure.

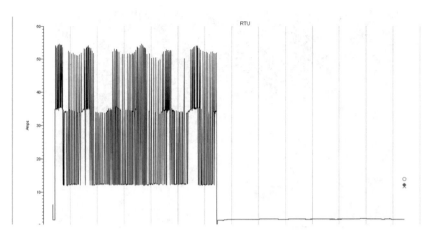

Figure 1-10. Logger Amperage

Energy Use Breakdown

The energy use breakdown for the plant is shown in Figure 1-12. The breakdown can be determined by actual measurements of individual energy-consuming equipment, like pumps, compressors, lighting and fans, or it can be calculated using the nameplate method, which uses capacity of the equipment and annual estimated operating hours. The nameplate method is less accurate.

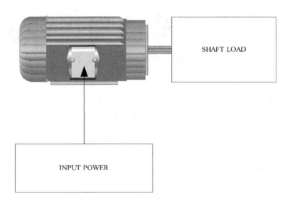

Figure 1-11(a). Input Power Measurements

Energy Analysis 11

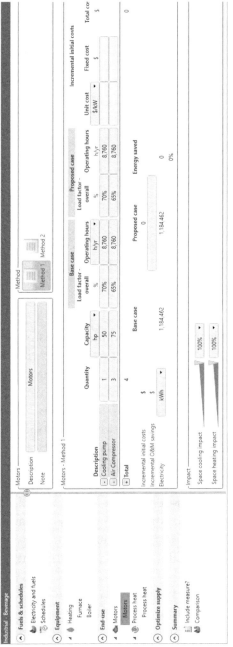

Figure 1-11(c). Nameplate method (RETSCREEN Expert—thermal energy)

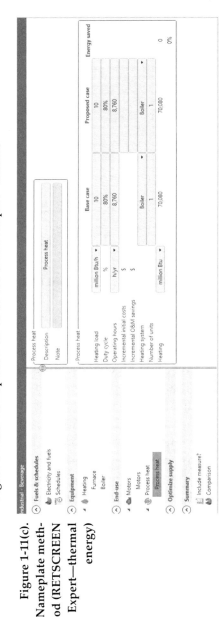

Figure 1-11(b). Nameplate method (RETSCREEN Expert—motors)

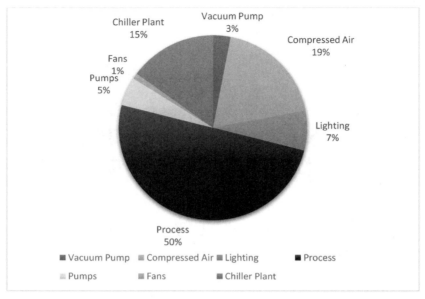

Figure 1-12. Energy Use Breakdown

Nameplate Method

In general, motors, lighting, electric heaters, and processing equipment consume the most electricity in manufacturing facilities. For electric heaters and processing equipment, the nameplate method uses rated power (kW) for all energy consuming equipment, multiplying the same by load factor to determine input power. The input power is multiplied by estimated annual operating hours to determine the annual energy consumption. For motors, the motor's horsepower can be obtained from the nameplate, and the operating hours would be provided by facility personnel. It is important to note here that the total annual energy for equipment should match the total annual bills. In a similar way, the annual energy consumption for heating equipment can be obtained from the capacity, load factor, and annual operating hours.

For gas consumption, the boiler or heater capacity, in Btu/hour or MBtu/hour, is multiplied by load factor to determine the demand, and consumption is calculated from estimated run hours.

Measurement Method

In the measurement method, larger equipment is logged using power analyzers or amp loggers. The logging should be conducted for a complete operating cycle. For smaller motors, spot amperage can be measured, and power can be calculated using the following formulae. Calculation of energy is based on power draw and estimated annual operating hours.

Similarly, gas consumption can be calculated from the load side. Some typical applications of thermal energy are the following:

- Air side sensible heat can be calculated from airflow and temperature difference. Air velocity can be measured using an anemometer or other air flow meter. Temperature can be measured by thermometers.

- Heating fluids or water. Thermal energy for thermic fluids or water is calculated from the specific heat of the fluid or water, mass of the heated fluid, and temperature differential of the fluid.

- From a steam flow meter installed in processing equipment, pressure and steam enthalpy can be used to calculate annual consumption in a process.

Energy/Process Flow Diagrams

Energy/process flow diagrams show material flows and other input to the process, such as compressed air, steam, chilled water, and others.

The energy flow diagram allows for a conceptual understanding of the energy and product flow, which can help with the following:

- understanding how different pieces of the production process are interlinked;

- identifying energy reduction opportunities, such as waste heat recovery;

- capturing various utility costs in the product flow process; and
- revealing optimization opportunities.

The following sections present an example energy process flow diagram from a plastic extrusion plant. In general, the manufacturing process consists of feeding raw material into screw extruders, where it is melted inside a barrel with electric heaters. The melted plastic is then passed through the molds. The profile is pulled by a puller sized per specifications, and cooled in a vacuum sizer. The plant uses chilled water for cooling, and the water in the plastic sheets is dried using compressed air before the sheets are cut according to specifications. Figure 1-13 provides a brief description of the production process.

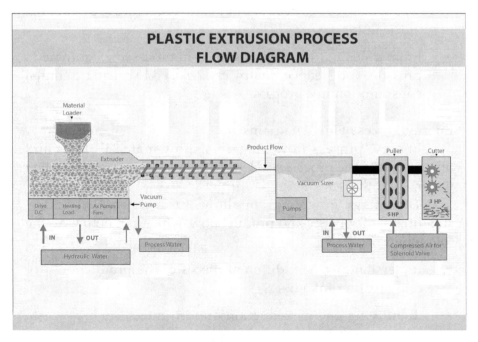

Figure 1-13. Plastic Extrusion Process

Energy Analysis

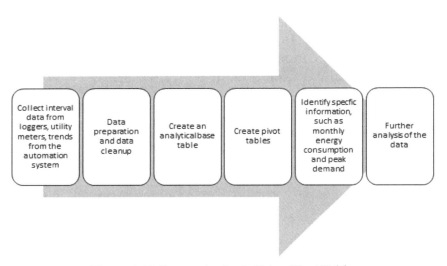

Figure 1-14. Energy Analysis Using Pivot Tables

Interval Data Analysis
Use of Pivot Tables in Excel for Data Analysis

Pivot tables help in organizing interval data on energy and related parameters, such as temperature, pressure, flow, production, etc., for analysis. The basic steps involved in creating a pivot are collecting interval data (from loggers, utility interval meters, etc.), preparing and compiling data, and creating pivot tables for specific analysis, such as monthly energy consumption and peak demand analyses. This information can be used for further analysis, such as creating regression models

The following example illustrates the use of pivot tables for interval data analysis. The example analysis intends to quantify energy consumption (kWh) and average demand (kW) for six days.

As a first step, the utility interval data are exported into Microsoft Excel. The data contain hourly kW information for eight days. The analytical base table is created using Excel functions YEAR (DATE), Month (DATE), and Day (DATE), to separate the year, month and day from the DATE values. The pivot table is formed from the analytical base table. The column in the analytical base table allows analysis across column parameters. In this example, creating a day column allows us to determine daily consumption for the data sample.

Interval data, or even logged data, can also be used to develop pivot tables. These pivot tables can then be used to develop the following:
- load profiles;
- review profiles for a specific period, such as a weekend or weekday or a month;
- base load review;
- peak load examination; and
- opportunities for energy savings due to implementation of a measure.

The above profiles can be useful to identify savings opportunities, such as idling equipment and management of peak demand for the facility. The profiles can also help in quantifying demand reduction due to implementation of energy management projects, such as lighting.

Table 1-1. Pivot Table

DATE	KW
2013-12-09 0:00	1421.64
2013-12-09 1:00	1419.48
2013-12-09 2:00	1454.04
2013-12-09 3:00	1413
2013-12-09 4:00	1434.6
2013-12-09 5:00	1416.96
2013-12-09 6:00	1422.72
2013-12-09 7:00	1957.32
2013-12-09 8:00	2052.36
2013-12-09 9:00	1978.92
2013-12-09 10:00	2131.2
2013-12-09 11:00	2117.88
2013-12-09 12:00	2034
2013-12-09 13:00	2097.72
2013-12-09 14:00	2145.6
2013-12-09 15:00	1873.44
2013-12-09 16:00	1923.12
2013-12-09 17:00	1797.12
2013-12-09 18:00	1883.88

The profiles in Figure 1-15 were developed using energy charting and matrix (ECAM) tools. The plant shuts down during the weekend. The peak load for the facility is 700 kW; it has a base load of about 450 kW during the week and a baseload of 300 kW on the weekend.

In the second example, the load profile for the facility is presented across several months. As shown, the interval data in-

Energy Analysis

Table 1-2. Pivot Table

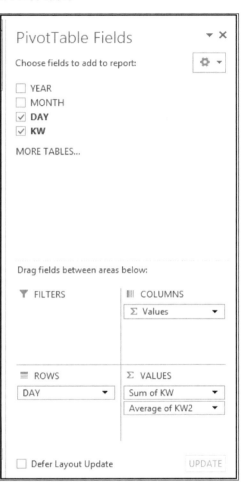

YEAR	MONTH	DAY	KW
2013	12	9	1421.64
2013	12	9	1419.48
2013	12	9	1454.04
2013	12	9	1413
2013	12	9	1434.6
2013	12	9	1416.96
2013	12	9	1422.72
2013	12	9	1957.32
2013	12	9	2052.36
2013	12	9	1978.92
2013	12	9	2131.2
2013	12	9	2117.88
2013	12	9	2034
2013	12	9	2097.72
2013	12	9	2145.6
2013	12	9	1873.44
2013	12	9	1923.12
2013	12	9	1797.12
2013	12	9	1883.88
2013	12	9	1803.96
2013	12	9	1745.64
2013	12	9	1823.04
2013	12	9	1730.16
2013	12	9	1479.24
2013	12	10	1429.2
2013	12	10	1450.44
2013	12	10	1488.6

dicate an increase in both peak load and base load for the facility. Based on a review of the load profile, a month-wise average load profile has been created from the filters. Comparison of the average load profiles during February and July is shown. Note the increase in the base load and in the peak load. The profile shown here is for weekday operation. Discussion with the operator revealed an increase in production that caused the plant to operate more equipment.

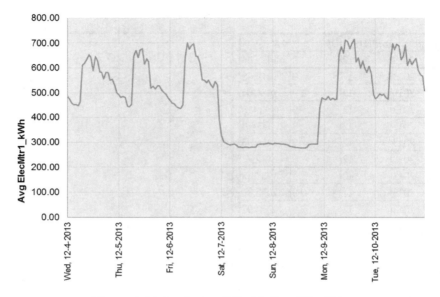

Figure 1-15. Analysis of Weekly Load Profiles

In another facility, interval data were reviewed, and the plant identified a higher base load on the weekend. Sub-meters were installed on some of the equipment; equipment idling was identified. The operator was trained to turn off equipment during the plant's non-operational hours. The interval data show the energy savings.

RETSCREEN Expert

RETSCREEN Expert is a clean energy management software for energy efficiency, cogeneration and renewable energy project. This tool has the capability in benchmarking feasibility analysis for projects and can be used for energy monitoring and targeting measurement and verification. RETSCREEN Expert has several data bases which include equipment details and specification, benchmarking data, cost data. It also connects to weather data from NASA. Application of RETSCREEN Expert is highlighted at various chapters in this book.

Energy Analysis 19

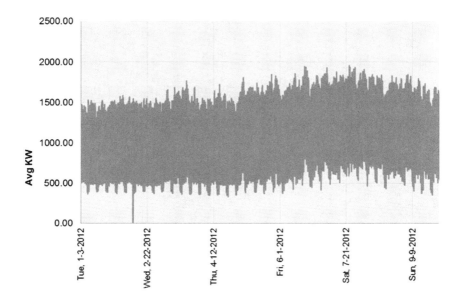

Figure 1-16. Monthly Load Profile Analysis

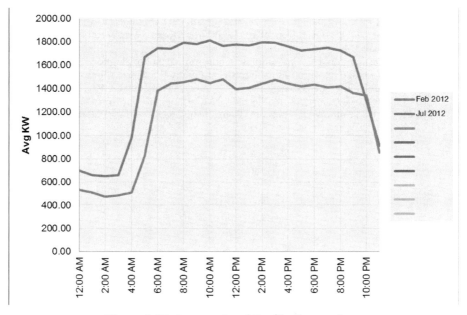

Figure 1-17. Average Load Profile Comparison

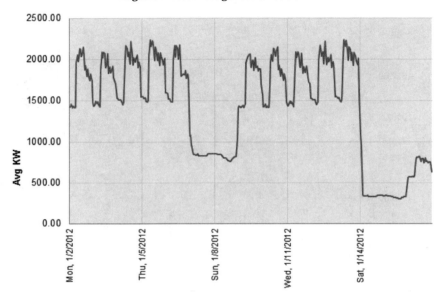

Figure 1-18. Average Load reduction

References

Taasevigen DJ and W Koran, 2013 User Guide to Energy Charting and Matrix Tool (ECAM), PNNL 21160, Pacific NorthWest National Laboratory, Richland, WA

NRCan. RETScreen. Retrieved from http://www.nrcan.gc.ca/energy/software-tools/7465

WEEC 2014,Washington, DC, Proceedings; "Short-term Data Logging to Identify Low-cost/No-cost Opportunities for Improving Energy Efficiency"; Tom White, P.E., CEM, Chief Engineer, Green Building Initiative; Ken Anderson, P.E., Principal, The Energy Gleaners; Paul Williamson, EMC, Principal, Planwest Partners; Kevin Stover, P.E., Commercial Programs Consultant, Green Building Initiative.

Energy Analysis

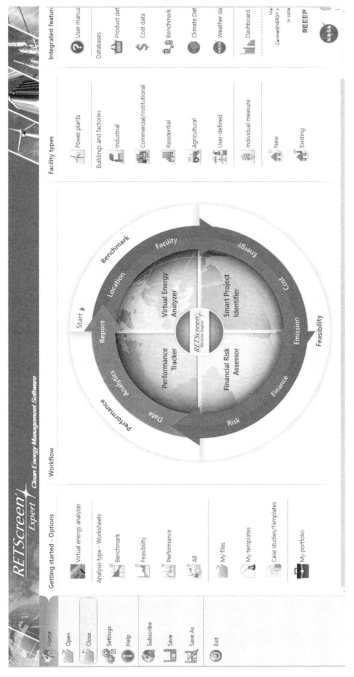

Figure 1-19. RETSCREEN Expert

Chapter 2

Energy Conservation Opportunities

Energy conservation opportunities arise from the following areas of waste:

1. Oversized equipment;
2. Equipment idling;
3. Matching supply and demand;
4. Improvement in operating efficiency;
5. Reduction of losses;
6. Waste heat recovery; and
7. System optimization.

A systematic approach is required to capture all energy reduction opportunities. The first step is understanding the system, for conceptual layouts are helpful. Questions on existing system operation should focus on matching supply and demand, part load operation, and operational issues supported by measurement. These questions will help identify areas of waste. The diagram in Figure 2-1 illustrates a general approach to identifying energy reduction measures.

This chapter discusses some of the common energy reduction opportunities found in the following systems: Industrial ventilation, industrial refrigeration, cooling water system, compressed air, fans, pumps and motors. The opportunities discussed in the following section are based on energy audits conducted on various industrial facilities.

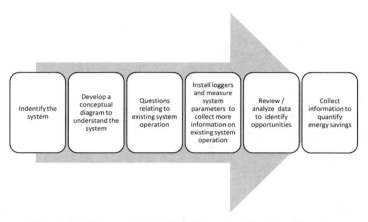

Figure 2-1. Energy Reduction Opportunity Identification

ENERGY CONSERVATION OPPORTUNITIES IN INDUSTRIAL VENTILATION SYSTEMS

Ventilation is key to providing a suitable working environment for employees and assists in maintaining safe operation of different processes in an industrial facility. It is one of the most overlooked areas in energy conservation, yet it can present several energy conservation opportunities, sometimes with minimal investment.

A typical industrial ventilation system includes various components, such as make-up air units, exhaust fans, and stacks. Exhaust fans exhaust contaminants from manufacturing facilities. Make-up air units provide ventilation air to a manufacturing plant.

Exhaust systems are designed for peak system operation when all equipment requiring exhaust is being operated. In most cases, all equipment is not operated simultaneously; therefore, the exhaust requirement is only a fraction of the installed capacity. Thus, some of the exhaust fans can be turned off; alternately, an exhaust system can be controlled by monitoring the plant's CO_2 levels. This also helps reduce make-up air intake.

This section discusses some of the energy conservation opportunities in industrial ventilation systems. These case studies are based on findings from industrial energy audits that were carried out in a food processing facility, a packaging facility, and a plastic extrusion plant located in Ontario, Canada.

The plants installed three make-up air (MUA) units of 3.8 MMBtu/hour each. The units are located on the roof. The MUA units have a capacity of 40,000 CFM each, and they are equipped with natural gas fired heaters and refrigeration compressors for summer cooling. The fan motor capacity of each of the make-up air units is 40 horsepower. The MUA system supplies treated air of 70°F during summer and winter according to plant personnel. The plant has installed several unit heaters controlled by space thermostats. There are a number of exhaust fans in the plant. The make-up air system, along with various exhaust fans, maintains positive pressure. The MUA system is operated seven days a week for 24 hours a day. An outside air sensor and discharge air sensor operate the burner. The heating in the MUA units is turned on whenever the outside air is lower than 62°F, while the cooling in the units was on when the outside air was above 80°F. Energy reduction opportunities in the MUA units were identified and are discussed below.

Shutdown MUA Units on Non-working Days

It was found during the energy audit that the MUA units operate continuously throughout the year. The MUA system can be easily shut down on Sundays and restarted on Mondays. It is important to ensure that while shutting off the MUA unit, the carbon monoxide level in the plant does not increase, so sensors should be installed to monitor carbon monoxide levels and operate the MUA unit whenever the carbon monoxide level increases beyond a set limit. This measure would reduce kWh consumption in the fan and natural gas consumption due to shutdown of the heating burner during winter. This would also reduce kWh consumption in the compressor and condenser fans during summer.

Temperature bin analysis was conducted to estimate gas and electricity savings. Heating and cooling energy is estimated for each temperature bin using Btu/hour = 1.08 X temperature difference X CFM. Since the cut-in temperature of heating mode is 62°F for the outside air temperature, and the cut-in temperature for cooling mode is 80°F, the temperature bin hours between 80 and 62 do not affect heating and cooling energy. The supply air temperature of 70°F was assumed for the analysis. By implementing this measure, the plant would reduce kWh consumption by 69,322 and gas consumption by 52,363 m^3 annually.

Improve Distribution System

Heating energy consumption can also be reduced by distributing heat directly to the point of use rather than throughout the entire facility. The distribution plenums for the roof-mounted MUA units direct the air into the plant in a generic pattern. As a result, the discharge temperature must be kept at 70 degrees to avoid comfort problems. Typical factory MUA is discharged at 60-65°F. Thus, a significant amount of heating energy does not directly contribute to heating the work area in winter. This is also responsible for creating hot and cold spots in the area. Improving

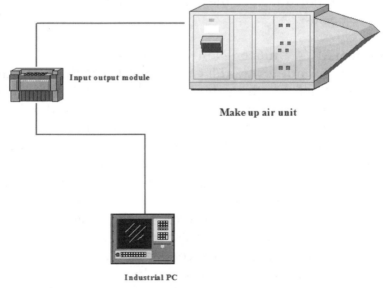

Figure 2-2. MUA Unit Control

Energy Conservation Opportunities

Table 2-1. Bin Analysis for MUA Unit

Temperature Bins TORONTO		Winter operation		Summer Operation		
		$(3)= 40000 * (70-(1))*$ $1.08 * 3$ (for three units)	$(4)=(3)*(2)$		$(6)=(5)*(2)$	$(7)=(6)*0.8$
Temp. (1)	hours (2)	BTUH (3)	BTU / YEAR (4)	Tonne (5)	Tonne hours (6)	kWh/year (7)
102	-					
97	1			150	150	120
92	11			150	1,650	1,320
87	52			150	7,800	6,240
82	184			130	23,957	19,166
77	328			-	-	-
72	440			-	-	-
67	659	-	-			
62	801	1,036,800	830,476,800			
57	712	1,684,800	1,199,577,600			
52	671	2,332,800	1,565,308,800			
47	552	2,980,800	1,645,401,600			
42	642	3,628,800	2,329,689,600			
37	862	4,276,800	3,686,601,600			
32	895	4,924,800	4,407,696,000			
32	530	4,924,800	2,610,144,000			
22	463	6,220,800	2,880,230,400			
17	393	6,868,800	2,699,438,400			
12	249	7,516,800	1,871,683,200			
7	161	8,164,800	1,314,532,800			
2	80	8,812,800	705,024,000			
-3	48	9,460,800	454,118,400			
-8	21	10,108,800	212,284,800			
-13	4	10,756,800	43,027,200			
-18	1	11,404,800	11,404,800			
Totals			28,598,430,000		33,557	26,845

Industrial Energy Management Strategies

Table 2-2. Energy Saving Calculation in MUA Unit

MAU system	Calculation	Unit
Annual gas consumption (from bin analysis)	28,598,430,000	BTU/year
Annual electricity consumption (from bin analysis)	26,845	kWh/year
Reduction of BTU due to shut off during Sundays *(28598430000 BTU/year * 0.065)*	1,858,897,950	BTU/year
Reduction of natural gas consumption due to shut off on Sundays *(1858897950 BTU/year/35500 BTU/m3)*	52,363	M3/year
Reduction of refrigeration electricity due to shut off on Sundays *(26845 kWh X 0.065)*	1,745	kWh/year
Fan power	40	HP
Operating Power of the Fan	18	kW
Annual kWh consumption for all 3 fans *(18kW * 3 fans * 24 hours/day * 365 days/year)*	473,040	kWh
Reduction of fan electricity due to shut off on Sundays *(Annual kWh consumption for all 3 fans * (1/7)) [Sunday are 1/7 part of a year]*	67,577	kWh

Total reduction in natural gas consumption	**52,363**	**M3/year**
Total reduction in electricity consumption	**69,322**	**kWh/year**

Energy Conservation Opportunities

distribution in the packaging area can occur by installing air bags to distribute the air.

Figure 2-3. Improved Distribution

In the north section of the plant, the existing system blows air directly at the hallway between packaging and production. It was recommended to install ductwork from the discharge plenums on the MUA units to about 8 feet above the production floor to deliver fresh air in winter and cooler air in summer directly to the workers. This would create better comfort and also allow a reduction in the MUA unit discharge air temperature setpoint in winter and an increase in winter. Bin analysis was conducted to calculate the energy savings due to reduction in temperature setting from 70°F to 65°F. By implementing this measure, the plant would reduce its natural gas consumption by 129,925 m^3.

Provide Insulation in the MUA Unit Duct

It was observed that the main air duct of the MUA unit did not contain any insulation.

Figure 2-4. MUA with Insulation

Insulation can reduce heat transfer in summer and winter. Bin data for Toronto were used to estimate the hourly reduction in heat loss if insulation was installed. To estimate the energy savings, the area of the uninsulated duct was measured to be 600 square feet. An R-value of 0.46 was assumed for the uninsulated duct and a value of three for 1 inch of internally insulated duct. Only heating energy reductions were considered. By applying insulation, the thermal energy reduction was calculated to be about 404,283 Btu/hour ft^2. By implementing this measure, the plant would reduce natural gas consumption by 6,833 m^3.

Optimize Energy in the MUA Unit and Exhaust System

This plant has installed several MUA units, exhaust fans, and about nine dust collection systems. The dust collection system picks up dust from the extruders (workstations) and transports the dust to the filters, where it is separated from the air.

Energy Conservation Opportunities

Table 2-3. Energy Saving Calculation, Insulation

Temperature (degree F)	Hours	Btu/hr·ft²	Btu/ft² (without insulation)	Btu/hr·ft²	Btu/ft² (with insulation)	Btu/ft² (Reductions)
(1)	(2)	(3)=(70−(1))/0.46	(4)=(2)*(3)	(5)=(70-(1))/3	(6)=(5)*(2)	(7)=(4)−(6)
102	0					
97	1					
92	11					
87	52					
82	184					
77	328					
72	440	0	0	0	0	0
67	659	0	0	0	0	0
62	801	17	13,930	3	2,136	11,794
57	712	28	20,122	4	3,085	17,036
52	671	39	26,257	6	4,026	22,231
47	552	50	27,600	8	4,232	23,368
42	642	61	39,078	9	5,992	33,086
37	862	72	61,839	11	9,482	52,357
32	895	83	73,935	13	11,337	62,598
32	530	83	43,783	13	6,713	37,069
22	463	104	48,313	16	7,408	40,905
17	393	115	45,280	18	6,943	38,337
12	249	126	31,396	19	4,814	26,582
7	161	137	22,050	21	3,381	18,669
2	80	148	11,826	23	1,813	10,013
-3	48	159	7,617	24	1,168	6,449
-8	21	170	3,561	26	546	3,015
-13	4	180	722	28	111	611
-18	1	191	191	29	29	162
Total thermal energy reduction						404,283

Install VFD on the Dust Collection System and Optimize Total System Air Flow

Each dust collection system is driven by a fan. The dust-collector fans operate continuously at 100 percent flow even if the extruders are not being used. The flow of the dust-collecting fan can be regulated based on the operational capacity of the system. Dampers/gates can be installed in the workstations. Each damper/gate in the extruders can be interlocked with the extruder's operation such that when it is not in operation, the gates are closed. The flow of the dust-collecting fan can be controlled to maintain constant pressure in the main duct. The other way of optimizing energy in the system is by permanently reducing the CFM in each dust collection system and matching the CFM to the existing operation. Assuming that, on average, 70 percent of the dust collection system is used, affinity laws can be used to determine the kW and energy kWh savings for the system.

Optimize Air Flow in the MUA Unit

The plant has a design air intake capacity of about 105,600

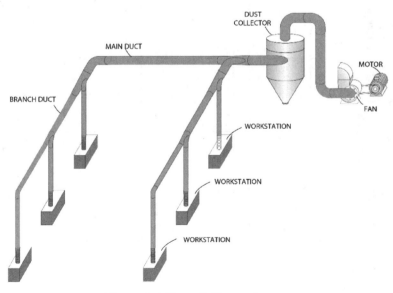

Figure 2-5. Dust Collector System

Energy Conservation Opportunities 33

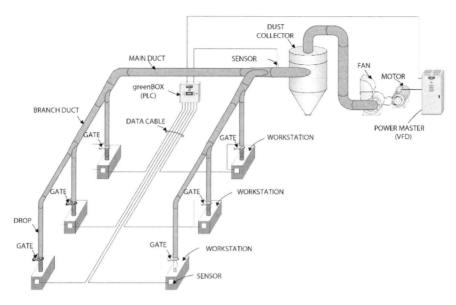

Figure 2-6. Optimized Dust Collector System (VFD)

CFM and total exhaust of about 222,781 CFM. During summer, the make-up and exhaust airflow rates are about equal, and little air is drawn into the plant through doors and openings in the building shell. During winter, more air is exhausted than is taken in. The dust collection system constitutes a large part of the air exhausted by the plant. Therefore, optimization of airflow in the dust collection system would reduce the quantity of air exhausted and also the make-up air. This would help in energy optimization, particularly during winter months when less make-up air heating would be required. Bin analysis was used to determine the energy savings, and it was estimated that by implementing the measure, the plant would reduce its natural gas use by 27,691 m^3.

The case studies show a host of energy conservation in the industrial ventilation system—some of which can be implemented easily with minimal investment, while others require investment but with quicker payback. With rising energy prices, the payback would be even shorter. There are other non-energy benefits of optimizing ventilation systems, like improved comfort levels, which should not be ignored in the analysis.

Table 2-4. Energy Saving Calculation Dust Collector

CAPACITY OF THE DUST COLLECTOR MOTOR (HP)	10
EFFICIENCY OF THE MOTOR	85%
INITIAL FLOW	100%
ANNUAL RUN HOURS	8,760
MOTOR LOADING	70%
INPUT POWER (KW) *(capacity of dust collector * 0.746 * motor loading)/(motor efficiency)*	6
ELECTRICITY CONSUMPTION	53,817
NEW FLOW	75%
NEW RUN HOURS	8,760
NEW KW *Using Affinity law, NEW kW = (Initial flow/new flow)^3 * input power*	2.59
ELECTRICITY CONSUMPTION	22,704
SAVINGS (KW)	3.55
SAVINGS (KWh)	31,113

CAPACITY OF THE DUST COLLECTOR MOTOR (HP)	15
EFFICIENCY OF THE MOTOR	85%
INITIAL FLOW	100%
ANNUAL RUN HOURS	8,760
MOTOR LOADING	70%
INPUT POWER (KW) *(capacity of dust collector * 0.746 * motor loading)/(motor efficiency)*	9
ELECTRICITY CONSUMPTION	80,726
NEW FLOW	75%
NEW RUN HOURS	8,760
NEW KW *Using Affinity law, NEW kW = (Initial flow/new flow)^3 * input power*	3.9
ELECTRICITY CONSUMPTION	34,056
SAVINGS (KW)	5.3
SAVINGS (KWh)	46,670

Energy Conservation Opportunities

Table 2-5. MUA Bin Analysis

Temperature Bins TORONTO		Winter operation	
temperature	hours	BTUH	BTU / YEAR
102	0		
97	1		
92	11		
87	52		
82	184		
77	328		
72	440		
67	659		
62	801		
57	712	19,819	14,111,185
52	671	52,851	35,462,940
47	552	85,883	47,407,239
42	642	118,914	76,343,096
37	862	151,946	130,977,693
32	895	184,978	165,555,382
32	530	184,978	98,038,382
22	463	251,042	116,232,298
17	393	284,073	111,640,878
12	249	317,105	78,959,215
7	161	350,137	56,372,070
2	80	383,169	30,653,510
3	48	416,201	19,977,633
8	21	449,232	9,433,882
13	4	482,264	1,929,057
18	1	515,296	515,296
		4,247,889	993,609,757

Table 2-6. Energy Saving Calculation

Volume of air heated by the MAU unit	105,600	CFM
New Volume of air heated by the MAU unit (due to reduction in air expelled)	99,483	CFM
Reduction in volume of air heated by the MAU	6,117	CFM
Heating temperature set point	60	F
Annual heating hours	Based on bin analysis	
Reduction in Heating energy requirement in the MAU	993,609,757	BTU
Reduction in Natural gas requirement in the MAU	27,989	Cubic meters

HEAT RECOVERY OPPORTUNITIES IN MANUFACTURING SYSTEMS

There are several heat recovery opportunities in an industrial plant. The heat recovery potential depends on the nature of process, operational strategy, equipment installed, and availability of suitable heat recovery sink.

Heat Recovery from a Compressed Air System

Heat generated from air compressors can be recovered and used for space heating application. The economic feasibility of such a measure depends on the hours of air compressor operation, the compressor capacity, and the location of the sink to capture waste heat. Installation cost can include heat exchangers, ducting, and other accessories, such as dampers and fans. The following case is taken from a food and beverages plant located in eastern Ontario. The plant had three screw compressors that operated about 8,000 hours per year. The compressors' capacity was 50 HP each. There were two make-up air units in the plant close to the compressors. The presence of a suitable sink close by made it a good candidate for recovering the waste heat for heating air in the MUA unit. The total energy available for space heating was estimated. This works out to be (0.80 x 150 bhp (for three 50 HP compressors) x 2,545 Btu/bhp x 4,000 hours/year) 1,221,600,000 Btu/year. Therefore, the natural gas savings due to heat recovery from air compressors would be 1,221,600,000 Btu/year/35500 Btu/m^3 = 34,412 m^3.

Blowdown Heat Recovery

In one energy audit, it was noted that steam boilers not returning all the condensate are good candidates for this type of conservation measure because make-up water often has contaminants, leading to high blowdown rates. The sink for this type of heat recovery method is generally the heating of make-up water. The following example is from a food and beverage plant in Ontario. The plant has a steam boiler rated for 400°F and 170 psi, with 3,520 lb/hour steam output. It has a maximum input of 4,148,000 Btu/

Energy Conservation Opportunities

hour and a minimum input of 1,382,000 Btu/hour. About 60 percent of the steam is used in the agglomerators. A portion is used in the dryers (15%), and a portion is used to heat water for a cleaning application a few hours per week. The steam boiler recovers only a small portion of condensate. The plant carries out intermittent blowdown in the steam boiler to remove the mud (particle) build-up in the boiler. It was proposed that the waste heat from blowdown be recovered by the boiler for preheating make-up water.

If a blowdown rate of 5 percent is assumed, the steam is blown down from a temperature of 350°F. A part of the total heat can be recovered from a temperature of 250°F to 100°F, and then the total heat recovered from blowdown would lead to natural gas savings estimated to be 3520 lb/hour × 5% blowdown rate × (250-100)°F × 8000 hours = 211,200,000 Btu/35,500, (with about 57% heat recovered) = 3,380 m³.

Figure 2-7. Boiler Blowdown System

Replace the Hot Water Heating System with a Direct Contact Hot Water Heater

In the above example, the steam is used for cleaning applications. Installation of a direct contact, gas-fired heater can help reduce energy consumption by improving thermal efficiency.

Figure 2-8. HE for Heating Water for Cleaning

The heating energy is supplied by the steam boiler through a heat exchanger; the efficiency of the present system is estimated at 60 percent, including boiler and distribution losses. By installing a direct contact hot water heater, it is possible to improve the thermal efficiency to about 95 percent. The plant installed a water meter to monitor the water consumption, which found the total hot water consumption for 10 days was 100 m^3. Assuming the final and initial temperatures of water were 175°F and 65°F, the heating energy required would be 8.34 lb/gal x 264 (conversion from m^3 to gallons) x 309 m^3/month x (175-65)°F x 12 months = 898,000,000 Btu. Therefore, the energy and cost reduction would be 898,000,000 x (1/0.6 – 1/0.95) = 551,000,000 Btu.

Heat Recovery from Process Reactors

Another common heat recovery method is combustion air preheating. This method uses the waste heat from the exhaust air to preheat combustion air. Many energy audits have shown that the hours of operation of the furnace being considered for combustion air preheating often affect the economics. The following case is taken from a chemical plant. The plant had five process reactors. The process chemicals and alcohol were heated in these reactors. The exhaust temperature of the reactor was maintained

Energy Conservation Opportunities

at about 1,500°F. It was proposed that the exhaust heat be captured by an air-to-air heat exchanger to preheat the combustion air. It was estimated that by implementing this measure, the plant could reduce its natural gas consumption by 9478 m³ annually.

Table 2-7. Heat Recovery Calculation

Item	Unit	Existing	Proposed
Volume of air	CFM	543	543
Air temperature	°F	90	180
Density of air	lb/ft³	0.075	0.03
Specific heat of air	Btu/lb °F	0.24	0.259
Operational hours			8760
Q = Volume of air per hour x Air temperature (difference)X Density of air X Annual Hours X specific heat (Average) X EFFYCIENCY HE			
Thermal energy reduction	Btu		336455435.6
Natural gas reductions	m³		9478

Implement a Condensing Economizer

The condensing economizer recovers heat from the boiler flue gases by cooling the exhaust gases below their dew point; the heat is often used for heating make-up water. Good candidates for this type of heat recovery method include large steam boilers that do not return all the condensate. The following case presents an example for the application of a condensing economizer. This case is taken from a chemical plant that has two steam boilers with a capacity of 450 horsepower each, supplying steam to various users at 125 psi pressure. The plant recovers only about 40 percent of the condensate. As a result, it is losing both water and energy content of the condensate. Even though they have installed a primary economizer in the boiler, there was still enough heat in the boiler exhaust that could be utilized. The plant was using a lot of make-up water but only returning 40 percent of the condensate. It was recommend that the plant recover the waste heat from the flue gas and use it to heat the make-up water using a condensing economizer. It was estimated that by implementing this measure, the plant would reduce the natural gas consumption by 273,802 m³.

Table 2-8. Energy Saving Calculation (Condensing Economizer)

Flue gas temp	311	°F
Make up water temp	60	°F
Make up water final temp	195	°F
Mass flow of the make up water	15,000	lb/hr
Hours	8,000	
% of condensate recovered	40%	
Boiler mass flow	9,000	lb/hr
Boiler efficiency	80%	
Heat recovered *Boiler mass flow rate x Make up water temp* *(difference)x hours of operation x efficiency of HE*	7,776,000,000	Btu/year
Fan blower capacity	15	HP
Electricity consumption in the blower	-53,712	kWh
Natural gas savings	273,803	M^3

PROCESS HEAT RECOVERY SYSTEM

This case presents an opportunity for recovering heat from a special process application. The case is taken from a coffee processing facility that has five roasters in the plant. Green beans are fed into the roasters, where they are roasted to a temperature of about 450°F. The exhaust from the roasters is then incinerated in the afterburners. Roasters 1, 2, 4, and 5 are batch roasters, while roaster 3 is a continuous roaster. The specific energy consumption of the coffee roasters varies widely from a maximum of 668 MJ/100 kg for roaster 1 to a value of 200 MJ/kg for roaster 3. The specific energy/operating cost of afterburners for different roasters along with the total annual operating costs are shown in Figures 2-9 and 2-10.

The total operating cost of the afterburners is about 65 percent of the total operating cost of the roasters. Calculations suggest that the operating cost of all roasters was $268,000, and the total cost of afterburners was $514,000. (The cost of afterburners increased in 2006 with the installation of a new unit.) Various en-

Energy Conservation Opportunities

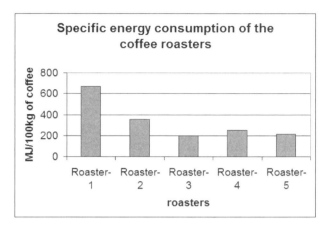

Figure 2-9. Specific Energy Consumption Coffee Roasters

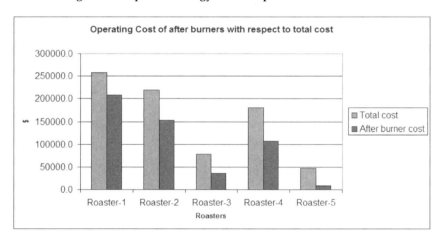

Figure 2-10. Operating Cost of Afterburner

ergy conservation options were studied. Below is a description of incinerator technologies.

Direct Fired Thermal Oxidizer

A direct fired thermal oxidizer (DFTO) is ideal for small airflow rates or batch processes that operate infrequently. The DFTO or current afterburner takes the roaster exhaust, heats the air to 1,400°F to destroy the odorous compounds, and then exhausts the clean air into the atmosphere. The equipment does not have a heat exchanger, which is why the operating fuel costs are so high. During the system operation, VOC-laden air will be exhausted into the system fan, through the burner section, and into the combustion chamber. The hot purified air will then be exhausted to the atmosphere.

Thermal Recuperative Oxidizer

The existing units could be rebuilt to add heat recovery to reduce operating costs. This heat exchanger addition would convert the equipment to what is referred to as a thermal recuperative oxidizer. With a thermal recuperative oxidizer, the odor-laden air from the roaster will be exhausted into the system fan and discharged into the primary heat exchanger, where it will be preheated. The odor-laden air will then move through the burner section and will be heated to the preset combustion chamber temperature. When the odor-laden air passes through the combustion chamber, the constituents in the exhaust creating the odor are oxidized, eliminating the odor. The hot purified air will then pass through the opposite side of the heat exchanger where it will preheat the incoming air. The purified air will then finally be exhausted into the atmosphere.

Catalytic Recuperative Oxidizer

During catalytic oxidizer operation, odor-laden air is exhausted into the system fan and discharged into the primary heat exchanger, where it can be preheated. After being heated on the tube side of the shell and tube heat exchanger, the odor-laden air will then move through the burner section. The burner adds heat as necessary to maintain the preset catalyst inlet temperature. When the odor-laden air passes through the catalyst, an exothermic reaction takes place. The hot purified air will then pass through the opposite side of the heat exchanger, where it will preheat the incoming air. The purified air will then be exhausted into the atmosphere. A precious metal catalyst is used to lower the oxidation temperature to 600°F. This compares to 1,400-1,500°F in a thermal oxidizer. A precious metal catalyst on a stainless steel monolith can be used to minimize catalyst volume and pressure drop.

Regenerative Thermal Oxidizer

A regenerative thermal oxidizer (RTO) consists of reinforced, insulated chambers filled with high temperature ceramic energy recovery media. A burner is located at the top of the RTO, between the two energy recovery chambers. The burner maintains the oxidizer above the oxidation temperature. Located beside the energy recovery chambers are diverter valves and air duct plenum passages which allow the process airflow to be diverted into and out of the oxidizer in either a clockwise or counter clockwise mode. The directional mode is controlled by a PLC, which changes the direction of airflow at regular intervals to optimize system efficiency. Typical operational cycles range from 2 to 4 minutes.

In operation, odor-laden air enters the oxidizer via an energy recovery chamber where the high temperature ceramic heat transfer media preheat the air prior to introduction into the oxidation chamber. As the air passes up through the bed, its temperature rapidly increases. After the oxidation reaction occurs, the hot, clean outgoing gas heats the exit energy recovery bed. To maintain the bed's optimum heat recovery efficiency, the airflow direction is switched at regular intervals by the automatic diverter valves on demand from the PLC control system. This periodic flow direction shift provides a uniform temperature distribution throughout the entire oxidizer. Although the illustration shows a forced draft (FD) fan, an induced draft (ID) fan is usually used for roaster emissions to keep oils and condensable emissions from collecting on the FD fan. To allow bakeout of oils that can accumulate in the coolest portions of the RTO, stainless steel internal components are often used.

The total annual cost reductions by implementing the alternatives are given in Table 2-9.

The estimated cost to retrofit each roaster with a 65 percent effective, 309 stainless steel shell and tube heat exchanger with a new fan is estimated to be approximately $140,000 to $230,000 for

Table 2-9. Cost Reduction Alternatives

Afterburners for Roasters	Operating Cost ($) of afterburners	Savings ($) (with HX)	saving ($) (with HX+catalyst)	Savings $ (with RTO)
Roaster 1	208,285	131,220	189,539	
Roaster 2	153,254	93,485	136,396	
Roaster 3	35,982	22,669	32,744	
Roaster 4	107,658	63,518	92,586	406,380
Roaster 5	8,908			
Total	514,087	310,891	451,265	406,380

equipment only. The cost to add heat recovery, a catalyst, and a new fan to each afterburner would be about $150,000 to $225,000. The equipment cost of a new RTO with an induced draft fan, bakeout, stainless steel internal cold-face materials, and 95 percent thermal energy recovery would be approximately $450,000. The heat recovery options in the example are capital intensive and would compete with other improvement opportunities in the plant. Fortunately, the plant was planning to upgrade the roasters already, so the heat recovery option was pursued as an opportunity to lower the operating cost of the coffee roasters. It was suggested that a life cycle analysis be conducted to determine the best alternative.

Conclusion

The heat recovery options presented above are well-known to most energy managers. The challenge is finding the applications that justify implementation. There have been examples where heat recovery options are being paid back in less than a year.

ENERGY OPTIMIZATION IN CHILLED WATER SYSTEMS

The major components of a chilled-water system include chiller, condenser pumps, chilled water pumps and cooling towers, while the components for an air-cooled chiller include chilled-water pumps and air-cooled condensers. The options to reduce chilled water costs in a manufacturing facility depend on certain variables, such as the system installed, type of chiller, chiller load profile, climate, economic factors, reliability issues, process requirement and application engineering issues. A techno-economic analysis of various energy conservation options applicable to chilled-water installation in a Canadian plastic extrusion plant was conducted.

The chilled water plant accounts for about 15 percent of the total electricity consumption in the plastic extrusion plant. It supplies cooling water to the extruders; specifically, the extrusion plant uses chilled water for cooling vacuum sizers. The centralized chilling plant has five 100-ton reciprocating compressors, for a total capacity of 500 tons. Chilled water for the vacuum sizers comes from a separate reservoir. Warm water is pumped from the reservoir into the chilling plant, where it is cooled and returned to the cold side of the reservoir. Warm water from the vacuum sizer cooling system is returned to the warm side of the reservoir.

COST-REDUCTION STRATEGY

Chiller Sequencing

The design objective of a chilling plant is to provide chilled water to the extrusion plant throughout the year at full-load operation. Therefore, the chilled water control point for each of the individual chillers was kept to low margins. However, due to operational changes, the demand for chilled water was reduced. This resulted in increased part-load operation of the chillers. The initial sequencing strategy of the chillers was not helpful in improving the part-load operation, which resulted in increased kW/TR in the system. Installation of proper sequencing controls can help reduce the kW/TR in the chiller banks.

Installation of Variable-flow Chilled and Condenser Water Pumping

The chilled and condenser water distribution loops are sized for five chilling machines; however, on average, three machines operate during the spring and fall. Optimizing the water flow in the chilled water pumps according to the operating tonnage of the machines by using a variable speed drive (VSD) would reduce the pumping energy in the system.

Installation of a Water-side Economizer

The water-side economizer system uses the cooling towers that normally cool the big water chillers, but it does so in a different way. Large heat exchangers, called "flat-plate" heat exchangers, are installed as part of the regular chilled water system. When conditions are right, such as in winter and during spring/fall evenings, it is possible to make chilled water using the cooling towers and the flat-plate heat exchanger, with the chillers off. Chillers consume a lot of energy, so cooling with them off is considered "free" cooling, even though some of the chilled water system motors are still running.

Conventional application of a water-side economizer calls for the chillers to be off and the cooling towers to make cold "condenser water," usually 40°F or so. By circulating the cooling tower water through one side of the flat-plate heat exchanger, the heat exchanger replaces the chiller, providing "free" chilled water on the other side of the heat exchanger. With several hours of temperatures lower than 40°F, an economizer would optimize energy in the chilling plant.

Chiller Bank Optimization

For the present application, the chiller bank optimization would involve installation of 300-ton VFD chiller with waterside economizer including VFD on the chilled-water pump and cooling-tower fan. The chilled-water control system would operate the plant to minimize the system kW/ton.

Installation of Energy-efficient Chillers

Chillers are the highest energy consumers in a plant; therefore, any improvement in their operating efficiency can lead to substantial reductions in energy consumption. Moreover, if the installed chillers are old, replacing them with more energy efficient chillers can reduce the operating cost of the system, even if resizing and optimization are not considered. In the present application it is proposed to replace the 100-ton (X5), reciprocating chillers with two 300-ton-VFD centrifugal chillers.

Implementation of Reset Controls

The condenser water reset operational strategy enhances efficiency of chiller operations as it reduces overall energy consumption in a chilled water system. The source of energy consumption in a chiller is the work required by the compressor where the refrigerant gas is raised from a lower pressure (evaporator) to a higher pressure (condenser). The effects of the increase in the refrigerant pressure differential between condenser and evaporator increase the work of the compressor.

Because the capacity of a cooling tower is a function of ambient conditions, lower ambient temperatures allow for additional cooling of the condenser water. The condenser water temperature decrease is a means to lower the pressure differential, thereby reducing the compressor's work. Reducing the temperature of the water entering the condenser (or the water leaving the cooling tower) by means of resetting increases the operational efficiency (kW per ton) of a chiller.

As a constraint to a condenser water reset strategy, a minimum pressure differential must be maintained between the condenser and evaporator for proper oil movement. Chiller efficiency may suffer if the cooling tower temperature is lowered to the extent that the return of refrigerant to the evaporator is impeded. Further, cooler condenser water temperature is obtained at a higher cost of energy consumed by the cooling tower fan.

Analysis Methodology and Application of Tools

Chilled water system analysis tools (CWSAT) were used to model the system and quantify the energy reduction in the chilling plant by implementing the energy conservation options discussed above. Location-specific weather files—in this case, for Toronto—were used to conduct the analysis (Figure 2-11).

Results and Discussion
Chiller Sequencing
CWSAT analysis shows that sequencing the chillers would result in a net reduction of 935,963 kWh in the chilling plant. This reduction is about 24 percent of the total chilling plant baseline energy consumption.

Installation of Variable Flow Chilled Water Pumping
Optimizing the water flow in the chilled water pumps according to operating tonnage of the machines using a variable speed drive (VSD) would reduce the energy in the chilled water system by 531,175 kWh. It would reduce the cooling tower energy by about 25 percent and pumping energy by about 28 percent. The reduction is about 14 percent of the baseline energy consumption. Relevant issues, such as maintaining minimum flow across the chillers need to be examined.

Installation of a Wet-side Economizer
CWSAT analysis shows that installation of a free cooling system would result in a net reduction of 278,241 kWh in the chilling plant. This reduction is about 7 percent of the total chilling plant baseline energy consumption. On the components level, there is a reduction of 19 percent in the chillers; however, there is an increase of 3 percent in the cooling towers.

The only catch to implementing the free cooling system is that once the flat plates can't keep up anymore, the system reverts to mechanical cooling. Therefore, issues related to the changeover from free cooling mode to chiller mode and its impact on the process need to examined carefully during implementation.

Chiller Bank Optimization
Optimizing chiller banks would reduce the electricity consumption in chillers by 2,519,402 kWh. The reduction is about 65

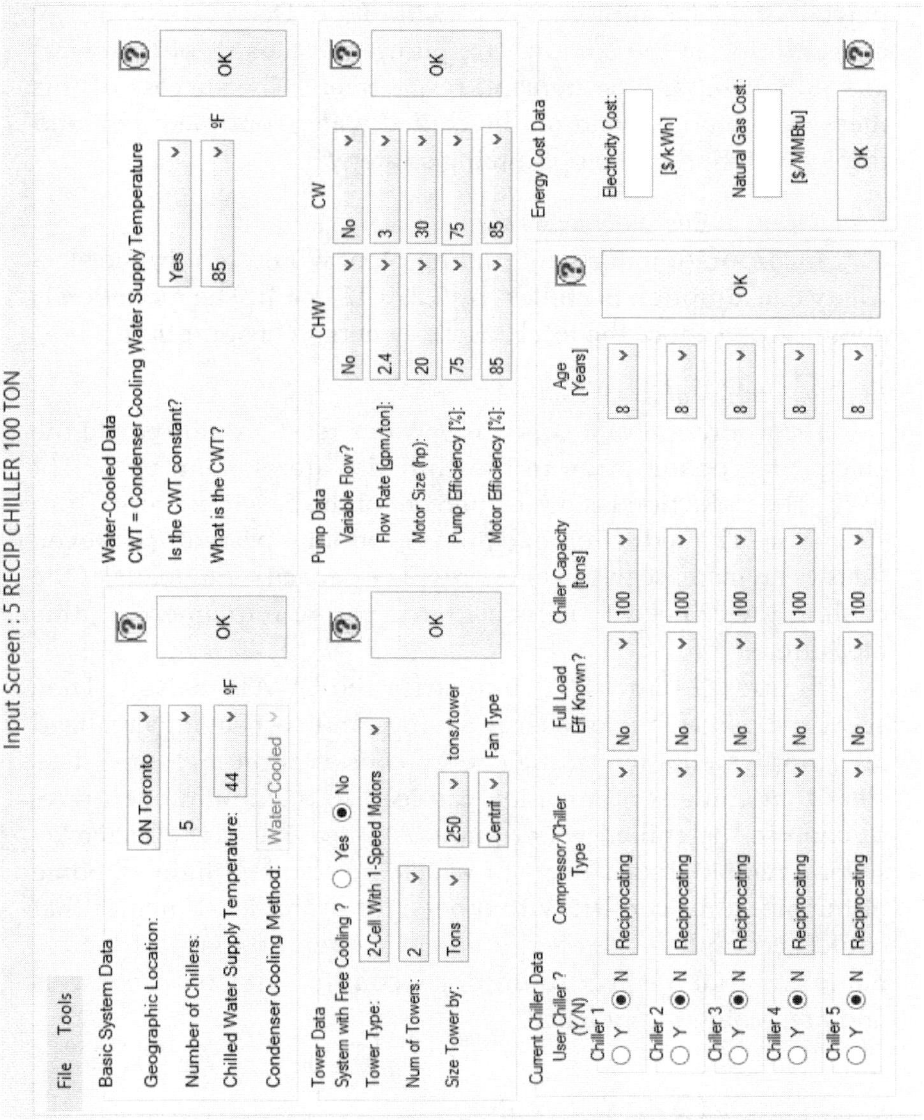

Energy Conservation Opportunities 49

Figure 2-11. Input Screen CWSAT

percent of the baseline energy consumption. Decreases in energy consumption in the compressor, pump and fans (cooling tower) are 56 percent and 84 percent, respectively. The success of this measure would depend on the chilled water control system and implementation of the optimizing strategy

Installation of Energy-efficient Chillers

Incorporation of a new chilling plant would reduce the electricity consumption in chillers by 2,299,525 kWh. The reduction is about 60 percent of the total baseline energy consumption.

Implementation of Reset Controls

Incorporation of a condenser water reset would reduce the electricity consumption in the chilled water system by 67,114 kWh. The reduction is about 2 percent of the baseline energy consumption. The reduction in chillers is about 6 percent. However, there is an increase in tower energy. The capacity and loads of the cooling tower need to be considered before implementing this measure.

Figures 2-4 and 2-12 summarize the CWAT analysis. They show the percentage of energy savings and cost savings realized after implementation of the energy conservation measures. It is clear that chiller optimization can lead to maximum energy reduction in the chilled water plant. All measures are affected by several related issues, some of which are also highlighted. Some of the measures conflict with others; for example, strategies that could accomplish the energy efficient operation at part load—namely, variable-speed pumping—conflict with reset controls and are avoided in practice.

ENERGY OPTIMIZATION IN AN INDUSTRIAL COOLING WATER SYSTEM

Cooling water systems typically consist of cold water tanks, generally known as a cold well; a hot water tank generally known as a hot well; a cooling tower; and pumps to distribute the water. The heat is picked up from the equipment being cooled and

Energy Conservation Opportunities

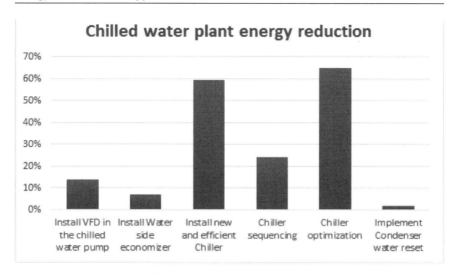

Figure 2-12. Percentage of energy reduction realized by implementation of the energy conservation options in the chilled water plant

stored in the hot well by one set of pumps. The heated water from the hot well is cooled in the cooling tower and stored in the cold well by another set of pumps. There are several opportunities to reduce energy in a cooling water system.

Optimizing Cooling Water Flow in Compressors

This case is taken from a metal processing facility where the cooling system supplies water to the compressors. Two circulating pumps (about 140 GPM) are operated for supplying water to these compressors. A discussion with the plant operator revealed that the same cooling system was supplying cooling water to several machines, including the compressors. However, the system use was modified to supply water only to the air compressors. The compressors' manufacturer was contacted to determine the cooling water requirements of the compressors. About 50 GPM is required to provide water to the cooling system. Therefore, it is recommended that the plant shut off one of the water pumps. Temperature-controlled valves can be installed in the compressor cooling circuit, and the flow of the operating pump can be regulated using a VFD (Figure 2-13). With this measure, the plant can reduce its energy consumption by 37,370 + 65,782 kWh.

Install VFD at the Circulating Pump

Another metal processing plant uses cooling water mainly in the annealers of the milling machines. Cooling for the milling machines is supplied by water from a central reservoir. As shown in Figure 2-14, warm water is pumped from the reservoir into a cooling tower, and cold water is returned to the cold side of the reservoir. One or more circulating pumps provides cold water to the main process loop in a constant flow. Water from these process loads is returned to the warm side of the reservoir. The loops are sized to accommodate three milling machines; however, according to an operator, on average, only about two machines operate at any one time. To handle the excess water being pumped through the loops, the supply and return headers at the ends of each loop are connected, allowing the excess water to return to the cooling tank without passing through the milling machines. It was recommended to close the "by-pass" valves at the ends of each loop and to install variable speed drives (VSD) on the 10-hp circulating water cooling loop motors. The VSD would control the pump speed to maintain a constant pressure drop between the supply and return headers. The VSD contractor should help locate where to install the differential pressure sensors between the supply and return headers and optimize control of the VSDs.

Reduce Valve Losses and Redesign Pumping System

In another cooling water system, there are seven cold well pumps used for transporting cooling water from the industrial cooling tower to evaporators 1, 2, and 3. Rated flow and pressures of individual pumps are 3,667 gal per minute and 219 FT, respectively. The pumps are driven by 250-HP motors. Five pumps are operated during summer and peak load conditions. Flow meters were installed on the common header to estimate the total water flow which comes to around 6,605 gal per minute.

It was observed that valves in all the pumps were throttled; discussion with plant personnel revealed that the throttling was done to prevent tripping of the driving motors, which indicates the motors might be undersized for the installed pumps. Input power measurement was carried out at all pumps. Details of

Energy Conservation Opportunities

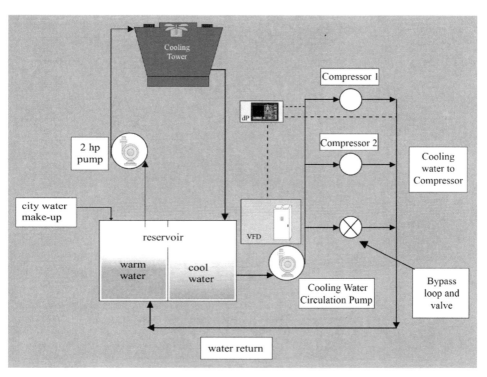

Figure 2-13. Compressor Cooling Water Optimization

Table 2-10.
Stopping one pump by reducing excess water flow in the compressor

Circulating Pump kW (before stopping one pump)	17
Circulating Pump kW (after stopping one pump)	7.9
kW Savings	9.1
Hours of Operation	7,200
Electricity Savings (kWh)	65,520
New baseline KW	7.9
Efficient case KW [using affinity laws old flow @100% and new flow @70%] =(Flow(old)/Flow(new))^3 * New baseline	2.7
Savings in demand due to VFD in the pump	5.2
Electricity Savings (kWh)	37,370

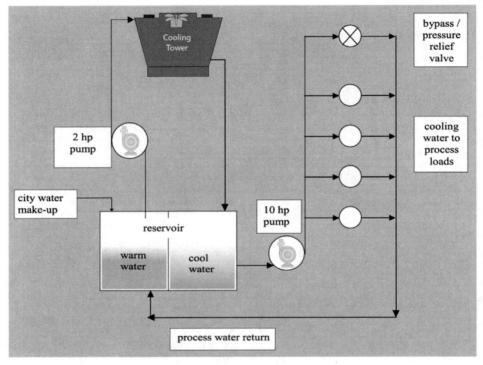

Figure 2-14. Process Cooling Water Loop

Table 2-11.
Estimate of Energy Savings after Installing a VFD in the Cooling Plant

Circulating Pump (kW)	4.5
Circulating Pump Flow (%)	100
Reduction in Pump Flow (%)	80
Efficient case KW [using affinity laws old flow @100% and new flow @70%] $=(Flow(old)/Flow(new))^3$ * New baseline	2.3
Electricity Savings (kW)	2.2
Annual Hours of Operation	7,200
Annual Electricity Savings (kWh)	15,811

measurements, including control valve position, are given in **Figure 2-12.**

According to the measurements, the present system is con-

Energy Conservation Opportunities

suming 795 kW power, and the possibility of replacing the existing pumping system with a new efficient system was explored. It was proposed to install a new efficient system consisting of two energy efficient pumps, which can cater to the flow and pressure needs of the plant. Hence, a new system of two pumps was proposed in which each pump will deliver a flow of 6,605 and a head of 197 ft.

ENERGY OPTIMIZATION IN INDUSTRIAL REFRIGERATION SYSTEM

Introduction

Industrial refrigeration can be found in many types of applications, such as food processing and product preservation. In many cases, industrial refrigeration represents one of the largest

Table 2-12. Operating parameters of cold well pumps

Pumps	% opening	kW
Cold well pump number 2	10	145
Cold well pump number 4	50	161
Cold well pump number 5	50	179
Cold well south	70	155
Cold well north	60	155
Total kW		795

Table 2-13.
Analysis for replacing the existing system with a new energy efficient system

Required flow	6605 gpm
Number of pumps	2
Head	197FT
Motor input power to meet the flow requirements (Flow * Head)/(3960 * pump efficiency * motor efficiency)	456 kW
Existing motor input power for five pumps	795 kW
Power savings	339 kW

consumers of energy in a facility. Studies show that reduction of energy consumption is possible in industrial refrigeration systems by as much as 40 percent.

The major types of equipment associated with industrial refrigeration cycle are evaporator coils that include fluid coolers and product coolers, and compressors. There are different types of compressors namely screw, reciprocating and rotary vane, and evaporative condensers or cooling tower.

This sections presents an analysis of the energy conservation opportunities identified in an industrial refrigeration system in a warehouse located in Buffalo, New York.

A breakdown of several of the facility's energy users is shown in Figure 2-15. It can be seen that refrigeration load accounted for the greatest amount of electricity consumption.

Refrigeration System

Four refrigeration circuits are installed at the facility: System A, System B, System C, and System D. Systems B, C, and D each have three reciprocating compressors, whereas System A has two compressors. Each system has its own evaporative condenser. Systems A, B and C serve various freezers in the facility. System D is a medium temperature loop and feeds four coils in the cooler, four coils in the loading dock, and one coil each in the vestibule (not operational) and tomato cooler. A central automation system in the facility controls the refrigeration system. The identified energy conservation measures are detailed below.

Energy Conservation Opportunities
Install Floating Head Pressure Control

The facility was maintaining a condensing pressure of 150 psig in Systems A, B, C and D. Reducing the condensing pressure would improve compressor efficiency and, therefore, optimize energy consumption in the compressors. In addition, lowering the pressure would put less stress on the reciprocating compressors and also increase compressor and system capacity. This control strategy would operate the refrigeration system with optimum

Energy Conservation Opportunities 57

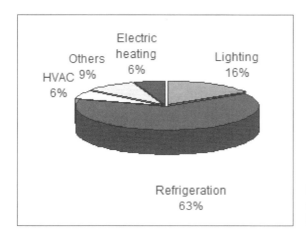

Figure 2-15. Breakdown of Annual Electricity Consumption

Figure 2-16. Refrigeration Condenser

condensing pressure based on outdoor wet bulb temperature (WBT). It was observed that each condenser fan and pump was cycled to control the discharge pressure. VFDs in the condenser fans to control discharge pressure are more efficient and would result in smoother operation with less wear and tear on the condensing tower motors, contactors, and belts. The automation system would have to be programmed to implement floating head pressure control. Additional points, such as discharge pressure sensors, need to be tied onto the automation system. It should be noted that compressor limitations, defrost, and other factors affect the minimum allowable discharge pressure, and the same would need to be considered when pursuing strategies to reduce the discharge pressure.

VFD Control of Fans in Evaporator Coils

In general, each evaporator coil has two or three fans of different horsepower levels. The evaporator coil capacity is controlled with the help of valves, while both evaporator fans operate even when the refrigerant flow to the coil is shut off. This form of control results in high energy consumption in the evaporators and also in the compressors. It was recommended that VFDs be installed in the evaporator fans, which would optimize the fans' energy consumption in the evaporators and also reduce the associated load of the compressors. The existing automation system could be utilized to implement VFD control of the evaporator fans.

Install More Efficient Refrigeration System for Systems B, C, and D

The refrigeration system employs only reciprocating compressors. A combination of screw compressors with reciprocating compressors would improve part-load operation. Redesigning Systems B, C, and D for a lower specific energy consumption by incorporating screw compressors with other elements that would reduce specific energy consumption in the system, such as incorporation of an economizer, can also be considered. Elements of the

Table 2-14. Floating Head Pressure Energy Reduction (for one compressor)

Dry Bulb Temperature	71.1	°F
Wet Bulb Temperature	57.2	°F
Measured Discharge temperature	78.2	PSIG
APPROACH (EVAPORATIVE CONDENSER)	21	EVAPORATIVE
MINIMUM DISCHARGE PRESSURE	150	PSIG
DISCHARGE TEMPERATURE AT MINIMUM DISCHARGE PRESSURE	84.3	°F
PROPOSED DISCHARGE PRESSURE	100	PSIG
DISCHARGE TEMPERATURE AT PROPOSED DISCHARGE PRESSURE	63.5	°F
PROPOSED APPROACH	15	
MOTOR HP	30	HP
MOTOR KW	19.9	KW

Energy Conservation Opportunities

Table 2-14. (*Cont'd*)
Floating Head Pressure Energy Reduction (for one compressor)

WBT (1)	HOURS (2)	CURRENT DISCHARGE TEMPERATURE (3) MAX(WBT + Approach $_{Evaporative\ Condenser}$, Discharge Temp)	PROPOSED DISCHARGE TEMPERATURE (4) MAX(WBT + Approach $_{Proposed}$, Discharge Temp)	Savings (5) ((3) - (4)) * 1%	KWh (6)
75.0	15.0	96.0	90.0	0.1	17.9
73.0	67.0	94.0	88.0	0.1	80.0
71.0	94.0	92.0	86.0	0.1	112.2
69.0	238.0	90.0	84.0	0.1	284.1
67.0	300.0	88.0	82.0	0.1	358.1
65.0	315.0	86.0	80.0	0.1	376.0
63.0	343.0	84.3	78.0	0.1	429.9
61.0	408.0	84.3	76.0	0.1	673.7
59.0	384.0	84.3	74.0	0.1	786.8
57.0	333.0	84.3	72.0	0.1	814.8
55.0	327.0	84.3	70.0	0.1	930.2
53.0	383.0	84.3	68.0	0.2	1,241.9
51.0	346.0	84.3	66.0	0.2	1,259.6
49.0	317.0	84.3	64.0	0.2	1,280.2
47.0	337.0	84.3	63.5	0.2	1,394.4
45.0	343.0	84.3	63.5	0.2	1,419.3
43.0	262.0	84.3	63.5	0.2	1,084.1
41.0	320.0	84.3	63.5	0.2	1,324.1
39.0	289.0	84.3	63.5	0.2	1,195.8
37.0	297.0	84.3	63.5	0.2	1,228.9
35.0	307.0	84.3	63.5	0.2	1,270.3
33.0	323.0	84.3	63.5	0.2	1,336.5
31.0	272.0	84.3	63.5	0.2	1,125.5
29.0	276.0	84.3	63.5	0.2	1,142.0
27.0	290.0	84.3	63.5	0.2	1,200.0
25.0	246.0	84.3	63.5	0.2	1,017.9
23.0	245.0	84.3	63.5	0.2	1,013.8
21.0	148.0	84.3	63.5	0.2	612.4
19.0	203.0	84.3	63.5	0.2	840.0
17.0	167.0	84.3	63.5	0.2	691.0
15.0	143.0	84.3	63.5	0.2	591.7
13.0	121.0	84.3	63.5	0.2	500.7
11.0	92.0	84.3	63.5	0.2	380.7
9.0	82.0	84.3	63.5	0.2	339.3
7.0	46.0	84.3	63.5	0.2	190.3
5.0	30.0	84.3	63.5	0.2	124.1
3.0	19.0	84.3	63.5	0.2	78.6
1.0	2.0	84.3	63.5	0.2	8.3
-1.0	5.0	84.3	63.5	0.2	20.7
-3.0	14.0	84.3	63.5	0.2	57.9
-5.0	9.0	84.3	63.5	0.2	37.2
-7.0	2.0	84.3	63.5	0.2	8.3
			TOTAL SAVINGS		28,879.21

existing system—such as reciprocating compressors, the accumulators, and the evaporators—could be utilized in the new system.

Operational Vestibule Coil

The evaporator coil in the vestibule area between the bakery and the bun freezer is fed by System D. The coil is not operational. It was also observed that the doors between the bun freezer and the vestibule and the door between the bakery and vestibule do not close properly. Air infiltration in the vestibule adds additional load on the bun freezer. Repairing the coil in the vestibule and reducing infiltration would transfer some of the additional load from the bun freezer to the vestibule. Because the bun freezer is fed by systems operating with much lower suction temperature, and thus a much lower EER, the reduction of load from the bun freezer to the vestibule would result in a net energy savings in the refrigeration system.

Replace 400W Metal Halide with
LED Lighting in the Cold Areas

The 400-watt metal halide lamps in the cold storage area can be retrofitted with LED lamps. This would reduce the lighting load in the area with lower lamp maintenance. Another benefit of the retrofit is reduction of the refrigeration load on the system. Energy saving calculation of the lighting retrofit is shown in Table 2-18.

Application of Heat Recovery

There are four evaporative condensers in the refrigeration system. The heat emitted by the individual evaporative condensers can be utilized for a space heating application. The plant can install ammonia-to-water heat exchangers to pick up the heat from the condenser water and feed the hot water to the unit heater. Additional pumps could be installed to circulate the hot water in the unit heaters. It should be noted that implementing floating head pressure controls would reduce the potential for heat recovery in the condensers as it operates with lower condensing temperature.

Energy Conservation Opportunities

Table 2-15. Saving Calculation Evaporator Fan Cycling

Evaporator fan cycling		
Evaporator Fan Capacity	73.5	HP
Present load factor of evaporator fans	95	%
Load factor of evaporator fans after fan cycling	75	%
Annual Hours of operation	7,200	
Load Reduction		
*Evaporator Fan Capacity * 0.746 * (load factor base case - load factor with fan cycling)*	11	kW
Energy Reduction	**78,957**	**kWh**

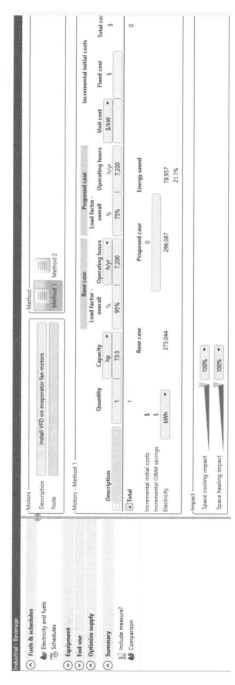

Figure 2-17. Saving Calculation Evaporator Fan Cycling

Table 2-16. Energy Reduction by Installation of Efficient System

Existing specific energy consumption of the compressor A	2.3	BHP/ton
Existing specific energy consumption of the compressor B	2.2	BHP/ton
Existing specific energy consumption of the compressor C	1.9	BHP/ton
Existing specific energy consumption of the compressor D	0.9	BHP/ton
Proposed specific energy consumption of the compressor A	2.1	BHP/ton
Proposed specific energy consumption of the compressor B	1.9	BHP/ton
Proposed specific energy consumption of the compressor C	1.6	BHP/ton
Proposed specific energy consumption of the compressor D	0.8	BHP/ton
Annual Average load on system A	60	Ton
Annual Average load on system B	42	Ton
Annual Average load on system C	26	Ton
Annual Average load on system D	8	Ton
Load Reduction $(BHP/TON_{existing\,A} - BHP/TON_{proposed\,A}) * Average\,Load\,A + (BHP/TON_{existing\,B} - BHP/TON_{proposed\,B}) * Average\,Load\,B + (BHP/TON_{existing\,C} - BHP/TON_{proposed\,C}) * Average\,Load\,C + (BHP/TON_{existing\,D} - BHP/TON_{proposed\,D}) * Average\,Load\,D$	26.9	kW
Hours	7,200	
Energy Reduction (kWh)	193,798	

Energy Conservation Opportunities 63

Table 2-17. Energy Reduction (Operationalize Coil)

Reduce infiltration and operationalize Vestibule coil		
Area of the room (Freezer)	10,000	sq feet
Height in the freezer	24	feet
Volume of the cooler	240,000	cubic feet
Air changes per hour	0.6	
Air Flow	2,400	CFM
H of outside air (80F, 50%RH) FROM PSYCRMETRIC CHART	31	btu/lb
H of air inside the freezer (-5F,90%RH) FROM PSYCHROMETRIC CHART	-0.6	btu/lb
Infiltration load		
$H_{outside\ air} - H_{inside\ air} * CFM * 4.5$	344,520	btu/hr
Refrigeration load	28.7	TR
Average load		
50% of Refrigeration load	14.4	TR
Existing specific energy consumption of the compressor A	2.3	BHP/ton
Existing specific energy consumption of the compressor B	0.9	BHP/ton
Annual Hours of operation	3,000	
Load Reduction	15	kW
Energy Reduction	**44,977**	**kWh**

Energy Efficiency of Pumps

Based on energy audits of pumps, it is often found that pumps, like motors, are oversized which results in low operating efficiency. The following section presents the background for calculating the operating efficiency of the pump and presents an example for them.

The output power of the motor, which is assumed to be equal to the input power to the pump (sometimes called the brake horsepower), is given by:

$$P_{\text{pump input}} = P_{\text{motor input efficiency}} \quad (1)$$

Output Power

The output power of the pump P_{out} (sometimes called the water horsepower) is equal to the rate of work done on the fluid, namely

$$P_{out} = yQhp \quad (2)$$

Where y is the specific weight of the fluid, Q is the volumetric flow rate, and hp is the energy increase per unit weight, or

Table 2-18. Calculation of Savings, Lighting Retrofit (Refrigeration)

Existing lamp type	Metal Halide	
Number	77	
Existing fixture wattage	450	W
kW of existing lamps	34.65	kW
Proposed lamp type	**LED**	
Number	77	
Proposed fixture wattage	165	W
kW of proposed lamps	12.705	kW
kW reduction	21.945	kW
COP of refrigeration system	2.5	
Hours of operation	7200	
kWh savings	158004	
kWh savings (with refrigeration effect)	63202	
Total savings	**221206**	**kWh**

Energy Conservation Opportunities

Table 2-19. Refrigeration Heat Recovery Calculation

Heat recovery from condenser		
Annual Average load on system A	60	Ton
Annual Average load on system B	42	Ton
Annual Average load on system C	26	Ton
Annual Average load on system D	8	Ton
Existing specific energy consumption of the compressor A	2.3	BHP/ton
Existing specific energy consumption of the compressor B	2.2	BHP/ton
Existing specific energy consumption of the compressor C	1.9	BHP/ton
Existing specific energy consumption of the compressor D	0.9	BHP/ton
Heat rejected by system = Average load on system * (12000 Btu/hr + Existing specific energy consumption * 2545)		
Heat rejected by system A	1,071,210	btu/hr
Heat rejected by system B	739,158	btu/hr
Heat rejected by system C	435,738	btu/hr
Heat rejected by system D	115,138	btu/hr
Total Heat rejection	2,361,244	btu/hr
% Heat recovered	50	
Hours	8,760	
Total Heat recovered	10,342,250,034	BTU
Natural gas savings (Heat recovered in BTU/35500)	291,331	M3

pump head, supplied to the fluid between inlet A and outlet B of the pump (Figure 2-18).

$$\left[\frac{p}{\gamma} + \frac{V^2}{2g} + z\right]_A + h_p = \left[\frac{p}{\gamma} + \frac{V^2}{2g} + z\right]_B$$

Since A and B are at the same reference height z_{ref}, this expression simplifies to

$$h_p = \frac{P_B - P_A}{\gamma} + \frac{V_2^2 - V_1^2}{2g}$$

where, in general, the entrance V1 and the exit velocity V2 have been assumed to have the same values.

Pump Efficiency

The efficiency of the pump is simply the ratio = P_{out}/P_{in}.

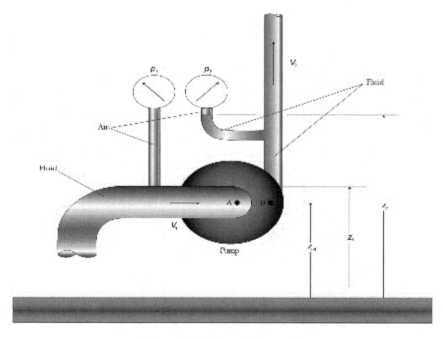

Figure 2-18. Pump Schematic

which is usually expressed as a percentage. This efficiency is found to be a function of the flow rate and is therefore not just a constant for a given pump. Efficiency is an important characteristic of a pump because, as indicated in Figure 2-19, the difference $P_{in} - P_{out}$ represents power that is lost.

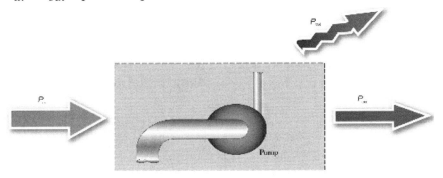

Figure 2-19. Schematic of Input and Output Power

Pump Characteristics

Pump manufacturers provide pump characteristic curves for their products. Fundamental variables appearing in pump characteristic curves are the input power P_{in}, flow rate Q, pump head hp, output power $P_{out} = Qhp$, and efficiency $= P_{out}/P_{in}$. Usually, flow rate Q is chosen as the independent variable, and the other quantities are plotted as functions of Q.

Manufacturers' curves are often quite compact, being simply plots of pump head hp as functions of Q, with efficiency appearing as a parameter along each curve. Only one such curve is needed to characterize a given pump completely. A sample calculation of a pumps efficiency is given in the next section.

Computation of the Efficiency of the Pumps

Data for flow measurement were noted down from the DCS; suction and discharge pressures were noted down from gauges installed in the field/DCS. Power measurements were determined by power analyzers and noted from meters installed in the field. Fluid properties, like specific gravity, viscosity, pumping temperature, etc., for specified pumps were given by plant personnel. Sample computation of pump efficiency for a cooling water pump is shown in Table 2-20.

Pump efficiencies were computed and compared with the respective design figures. It is recommended to operate the pump at flows close to the BEP because the operating cost and the loads on the other rolling elements becomes minimal at this point.

Compressed Air System

A compressed air system typically consists of air compressors, distribution piping, a storage tank, and driers.

Case Studies

There are five air compressors in the plant. Three of them are 75 HP lubricated screw compressors, while two are vane compressors, 50 HP and 30 HP. The vane compressors are operated as standby compressors and are not considered for calculating savings. Each screw compressor has its own refrigerated air dryer. The compressed air is collected in a receiver before it goes to the plant. Amperage data on all the operating air compressors—namely, compressors 4, 5, and 3—were logged to determine the load profile.

Based on the amp load profile, Air Master (US Department of energy tool) was used to determine an approximate hourly operating ACFM. Compressor 3 was not delivering any air, though it was still drawing amps.

Based on field measurements and site observations, the following potential energy conservation measures were identified: Install compressor sequencer and reduce leakage.

Install Sequencer

At present, all compressors have the load and unload pressure set at the same level, so more than the required number of compressor banks are online even when the air demand is not present. Based on the logged amps and generic compressed air characteristic curves, it was observed that one of the compressors (compressor 3) was idling without delivering air to the system.

Energy Conservation Opportunities 69

1	Motor Rating	75	HP
2	RPM	1,800.0	
3	Pump Flow	1,140.0	GPM
4	Input power	53.8	KW
5	Motor Efficiency	75%	
6	BHP (Motor efficiency x input power)/0.746	54.1	
7	Difference between inlet and outlet pressures from pressure gauge	22.0	PSI
8	Specific Gravity of water	1.0	
9	Head in FT (2.31X DIFFERTIAL PRESSURE (22.0))/ (Specific gravity =1)	50.8	Feet
10	Pump WHP [(PUMP FLOW X HEAD)/3960]	14.6	HP
11	Pump Efficiency (Pump WHP= 14.6)/ (MOTOR BHP= 54.1)	27%	

MOTOR RATING	25.0	HP
RPM	1,800.0	RPM
New Pump Efficiency	80%	
New Motor Efficiency	90%	
Pump (Water HP)	14.6	HP
Pump input HP (Water HP)/(Pump efficiency)	18.3	HP
Motor input power (Pump input power / Motor efficiency)	20.3	HP
Motor input power	15.2	kW
Savings (Input power (base) - input power (efficient))	38.6	kW
Annual Hours of operation	8,000.0	
Energy Savings	308,962.4	

Table 2-20. Energy Reduction, Pump

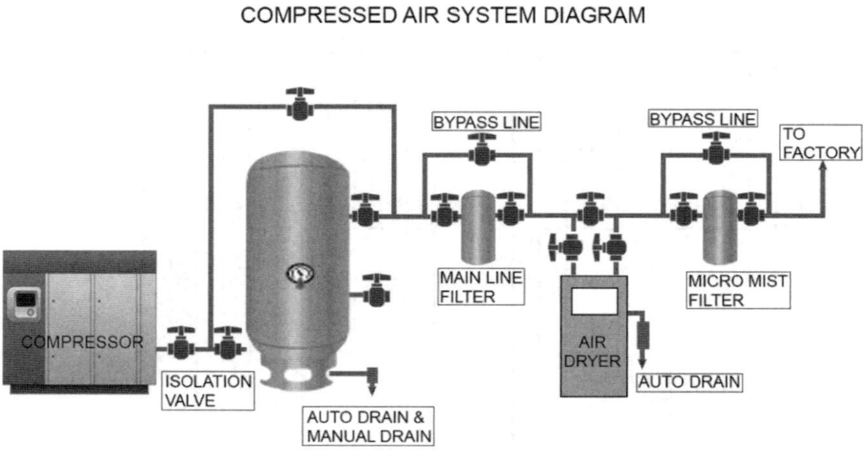

COMPRESSED AIR SYSTEM DIAGRAM

Figure 2-20. Compressed Air System

Installing a sequencer and sequencing the compressors based on air flow would shut off the idling compressor. Sequencing would also improve the kW/CFM of the compressors from 0.2 to 0.19 due to optimal loading. Also, energy consumption would be reduced due to the narrower pressure band. It was mentioned during the field visit that the minimum pressure required to operate various machines is 85 psi. The pressure can be lowered from an average of 100 psi to about 90 psi when the compressors are properly sequenced.

Install VFD compressors and Reduce Air Leakage

A food processing plant had two 100 HP screw compressors. The entire demand of the plant was served by running one 100 HP compressor at full load; the other 100 HP compressor was used to supply demand for 10 hours a day. The compressor supplied air at 100 psig. The leakage rate in the compressed air line was estimated to be 20 percent of the total air production capacity, a conservative figure considering the plant did not have a leak reduction program and most of the distribution system consisted of screwed schedule-40 steel pipe. The assumption was that the leakage can be reduced from 20 percent to 10 percent. It was pro-

Energy Conservation Opportunities 71

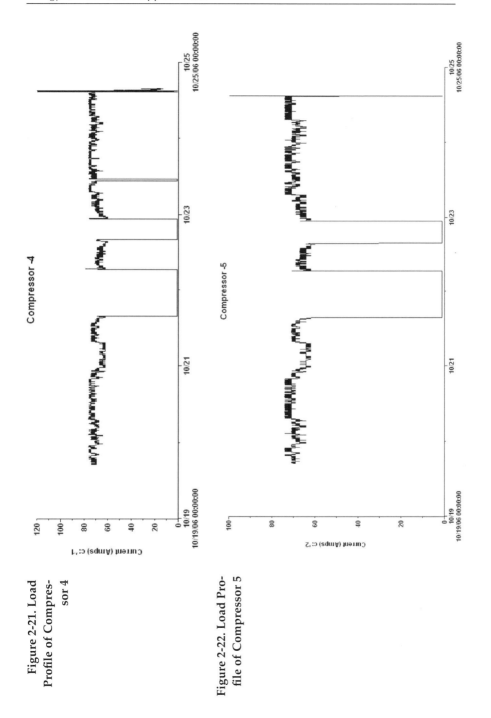

Figure 2-21. Load Profile of Compressor 4

Figure 2-22. Load Profile of Compressor 5

72   Industrial Energy Management Strategies

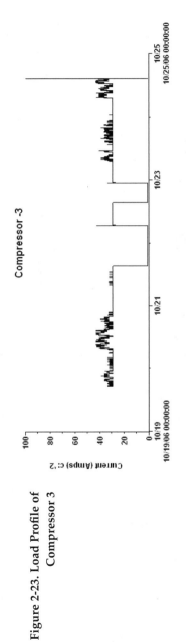

Figure 2-23. Load Profile of Compressor 3

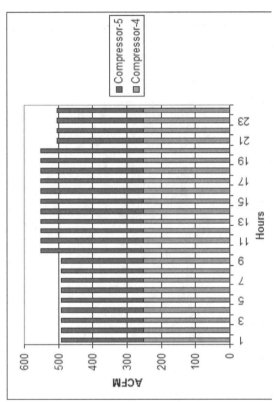

Figure 2-24. ACFM of Compressors 4 and 5

Energy Conservation Opportunities

posed to install VFD. RETSCREEN EXPERT was used to calculate the savings.

While large leakages are easily detected by sound, it is difficult to locate small leakages, which can only be detected by applying soap solution or by ultrasonic leak-finding equipment. Some of the most susceptible points are threaded pipe joints,

Table 2-21. Energy Savings due to Compressor Sequencing

Factor	Measurement	Unit
SAVING DUE TO IDLING		
Average kW consumption of the idling compressors	24	kW
Idling time for a year	6,000	hours
kWh saving due to shutting of air compressors	144,000	kWh
SAVING DUE TO COMPRESSORS SEQUENCING		
Specific energy consumption of compressors (present)	0.2	kW/CFM
Specific energy consumption of compressors (after)	0.19	kW/CFM
Difference	0.01	kW/CFM
Average capacity of the compressors	521	CFM
Demand reduction (kW)	5.21	kW
Annual hours	6,000	hours
Electricity reduction due to sequencing	31,260	kwh
SAVING DUE TO PRESSURE REDUCTION (See chapter on operational savings)		
Proposed discharge pressure (90+14.7)	104.70	PSI
Current discharge pressure (100+14.7)	114.70	PSI
Proposed pressure factor = K = ((104/14.7)^0.285) -1	0.75	
Existing pressure factor = L =((114.7/14.7)^0.285) -1	0.80	
K/L	0.94	
Savings (D) = (1- 0.94)	0.06	
Average air flow	521	CFM
kW/CFM	0.20	kW/CFM
Reduction of kW due to pressure reduction	6.03	
Run hours	6,000	
Electricity savings	36,202	kWh
TOTAL SAVINGS		
Total demand reduction kW (24+5.21+6.03)	35.24	kW
Total energy savings (36202.51+31260+144,000) kWh	211,462	kWh

flange connection valves, steam traps and drains, filters, hoses, connectors, operative valves on pneumatic devices, check valves, relief valves, and end-use machines or tools.

A proper leak reduction program entails finding, tagging, and recording leaks, and then ensuring leaks are promptly fixed. A pressure decay test should be done on a monthly basis so leaks can be identified promptly. Installing an air meter with a high turn-down ratio permits the plant to quantify leakage.

Additional steps that can be taken to reduce leakage include making sure that valves are closed properly while the machines are not in operation and fixing any leaks immediately.

References

New Saving Opportunity: On-demand Industrial Ventilation; Ales Litomisky, *Energy Engineering* Vol. 103, No. 3, 2006.

Energy conservation through improved industrial ventilation in small and mid-sized industrial plants, Saman, N.F.; Nutter, D.W. 16th Annual Industry conference 1994.

Bhattacharjee, Kaushik (2009). Energy conservation opportunities in heat recovery system, presented and published at World Engineering Congress Conference, Washington, DC, 2009.

Bhattacharjee, Kaushik (2009). Energy conservation opportunities in an industrial refrigeration system. *Journal of the Association of Energy Engineers*. 106 (3), 73-79.

Bhattacharjee, Kaushik (2008). Energy conservation opportunities in industrial ventilation system, presented at the World Engineering Congress Conference, Washington, DC, 2008.

Bhattacharjee, Kaushik (2006). Reducing energy cost in an industrial chilled water plant, presented at the World Engineering Congress Conference, Washington DC, 2006.

Bhattacharjee, Kaushik (2007). Energy conservation opportunities in an industrial refrigeration system, presented at the World Engineering Congress Conference, Atlanta, Georgia, 2007.

Bhattacharjee, Kaushik (2008). Energy conservation opportunities in an industrial refrigeration system. *Journal of the Association of Energy Engineers*. 105 (5), 55-63.

Bhattacharjee, Kaushik (2007). Reducing energy cost in an industrial chilled water plant. *Journal of the Association of Energy Engineers*. 104 (4), 42-51.

http://www.mass.gov/anf/docs/dcam/mafma/manuals/chilled-water-systems-analysis-tool.

https://energy.gov/eere/amo/software-tools

http://eeref.engr.oregonstate.edu/Opportunity_Templates

Thumann, A; Paul Mehta, D. (year). *Handbook of Energy Engineering*. The Fairmont Press, Inc. Lilburn, Georgia.

Energy Conservation Opportunities

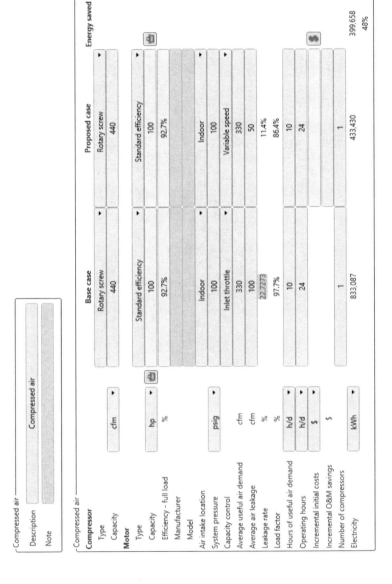

Figure 2-25. Energy Saving by Installing VFD Compressor

Chapter 3

Lean Manufacturing Principles And Their Impact on Process Energy Efficiency

Michael Stowe, P.E., C.E.M, P.E.M., CP EnMS

INTRODUCTION

Background

The benefits of implementing lean manufacturing are well documented. Lean manufacturing practices can have a very positive impact on industrial plant process efficiency and the associated financial bottom line.

Andon is a lean manufacturing term that refers to a system to notify team leaders, managers, maintenance technicians or other workers of a quality, material or process problem at a discreet piece of processing equipment. Specialized equipment is available that can measure, record, and document lean characteristics of industrial process equipment and then communicate this information to a dashboard or a specific employee via alerts.

This chapter documents some key lean manufacturing principles and then relates these to various available Andon hardware and software that can be used for tracking and continuous improvement. There are both energy and non-energy benefits to using Andon related equipment.

Objective

The objective of this chapter is to provide a summary of some key lean principles and related Andon hardware, software, and systems that can be implemented to document lean processes for the benefit of the applicable industrial production line.

Applications, Value and Use

This chapter is oriented towards industrial end users who are manufacturing products through transformation processes. It is intended to help energy professionals assist their industrial customers with understanding lean manufacturing measurement equipment and the associated energy and non-energy benefits. It can also serve as a brief overview of lean manufacturing. Finally, lean manufacturing and its relationship to energy efficiency are also explored.

LEAN MANUFACTURING BASICS

What is Lean?

Lean manufacturing, often just referred to as "lean," is a production philosophy that considers the expenditure of resources for anything other than the direct creation of value for the end customer to be wasteful, and therefore a target for elimination. Working from the perspective of the client who consumes a product or service, "value" is any action or process that a customer would be willing to pay for. Essentially, lean is centered on making obvious what adds value by reducing everything else [1].

In other words, if the process or action is not adding value that the end customer is willing to pay for, then the process or action is a form of waste. The customer is not willing to pay for scrap or for parts sitting around waiting for processing or for moving the part from the machine to the warehouse. These are all forms of waste and must be eliminated or minimized. Essentially, lean, as the name implies is the trimming away of all the fat (i.e. waste), which leaves in place an optimized process made up of value adding essential actions only.

There are tremendous resources and details on lean manufacturing terms, systems and tools available on the internet [2, 3, 4]. For the purposes of this chapter, there are four key interrelated

Lean Manufacturing Principles and Their Impact on Process Energy Efficiency 79

areas of lean manufacturing that we will focus on because they relate to equipment and systems that can measure, record, and document lean characteristics of industrial process equipment.

These four key areas are:
- Andon
- Muda
- Overall Equipment Effectiveness (OEE)
- Kaizen

Andon

Andon, literally the Japanese word for paper lantern, is a manufacturing term referring to a system to notify team leaders, managers, maintenance technicians or other workers of a quality, material or process problem. The centerpiece is a signboard incorporating signal lights to indicate which workstation has a problem. The alert can be activated manually by a worker using a pull cord or button, or may be activated automatically by the production equipment itself. The system may include a means to stop production so the issue can be corrected. Some modern alert systems incorporate audio alarms, text, or other displays [5].

Andon is a visual system that frequently incorporates color coded lights and real time communication displays. A key feature of an Andon system is that it brings immediate attention to process problems right when they occur and notifies the correct personnel who help to correct the problem. The use of Andon will be the key focus of this chapter.

Here is an example of an Andon system. An operator notices that he is about to run out of a part so he presses a blue Andon button which signals a blue light indicating a material request. A material associate then responds to the signal and the part is replenished before the bin runs out. This keeps the process moving without having any downtime.

Muda

Eliminating waste is the central theme of lean manufacturing. Muda is the Japanese word meaning uselessness, idleness, or literally waste [6]. Muda consumes resources, but adds no value. Muda takes many forms, but is commonly categorized in lean manufacturing practices into seven areas, plus one, as follows:

Seven Areas of Muda:
- Transportation—Moving parts from one operation to the next
- Inventory—Raw materials, work-in-process, or finished products that have not yet produced any income
- Motion—Unnecessary wear and tear or accidents on equipment and employees
- Waiting—Parts sitting in front of a machine waiting to be processed
- Over-Processing—More work is done on a work piece than the customer has paid for
- Over-Production—More items are produced than the customer needs at a given time
- Defects—The item does not meet the customer's specification and requires rework or replacement, which is very costly

Plus one:
- Ineffective use of employee talent—Not taking advantage of the full range of creative skills of employees, which detracts from continuous improvement

Elimination of these wastes can be assisted with Andon type systems because immediate notification of problems is available to key personnel. This relationship will be discussed in more detail in the equipment section of this chapter.

Overall Equipment Effectiveness
Overall Equipment Effectiveness, known as OEE, is a methodology to measure how effectively a specific machine or a production line is utilized. OEE is made up of the product of three factors:
- Availability

- Performance
- Quality

Each factor is measured as a percentage with a maximum of 100 percent each. The factors are reduced from 100 percent by losses and wastes. It is nearly impossible for any manufacturing process to run at 100 percent OEE. A common target level for world class OEE is 85 percent. More details on these three factors are below [7, 8].

- Availability = {Actual Operating Time / Scheduled Operating Time} x 100
  — This represents the percentage of scheduled time that the operation is actually available to operate
  — The availability factor is a pure measurement of uptime and is designed to exclude the effects of quality, performance, and scheduled planned downtime
  — Major impacts on availability include equipment breakdowns and changeovers
  — If a machine was scheduled to run 450 minutes in an eight hour shift and due to a breakdown was only able to run 400 minutes, then the availability would be:
- Availability = (400 minutes / 450 minutes) x 100 = 88.9 percent

- Performance = {(Parts Produced x Ideal Cycle Time) / Actual Operating Time} x 100
  — This represents the speed at which the machine or process runs as a percentage of its designed speed
  — The performance factor is a pure measurement of speed and is designed to exclude the effects of quality and availability
  — Performance is a measure of how well the equipment performed when it was running
  — Major impacts on performance include minor stops, time for adjustments, speed loss and machine issues requiring

slower feed or cutting rates
— If a machine has an ideal cycle time of 40 parts per hour or 1.5 minutes per part and produced 248 total parts in the actual operating time of 400 minutes, then the performance would be:

- Performance = {(248 parts x 1.5 minutes/part)/400 minutes} x 100 = 93.0 percent

- Quality = {(Parts Produced – Defective Parts)/Parts Produced} x 100
   — This represents the good parts produced as a percentage of the total parts produced
   — The quality factor is a pure measurement of process yield and is designed to exclude the effects of availability and performance
   — Producing defective parts is a loss and a waste
   — The major impact on quality is producing defective parts that do not meet the customer's specifications and must be reworked or scrapped
   — If a machine produced 248 parts during a shift and 11 of these parts are defective, then the quality would be:

- Quality = {(248 parts – 11 defective parts)/248 parts} x 100 = 95.6 percent
- The equation to now calculate the OEE is:
- OEE = Availability x Performance x Quality

For our three factor examples above:
- OEE = (88.9%) x (93.0%) x (95.6%) = 79%

Figure 3-1 is a graphic representation of the OEE concept. Each subsequent pillar shows the impact of the losses in availability, performance and quality [9]. Notice how short the blue "Fully Productive Time" column is compared with the blue "Plant Operating Time" column. The difference is non-productive time.

Lean Manufacturing Principles and Their Impact on Process Energy Efficiency 83

OEE is an excellent indicator of the level of efficient utilization for a machine or process line. The factors that impact OEE are many of the same seven Muda wastes discussed earlier. Again, elimination of these wastes can be assisted with Andon type systems because the immediate notification of a problem is available to key personnel. Having some type of Andon system can help eliminate waste and improve OEE.

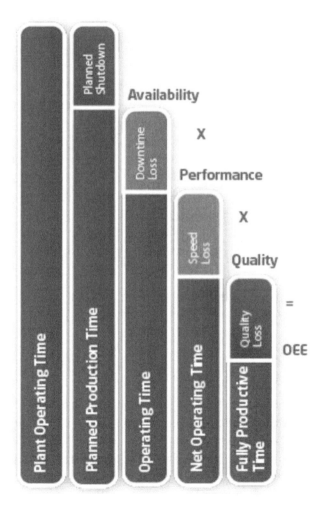

Figure 3-1. Graphic Representation of OEE

Kaizen

As mentioned already, the main purpose of lean manufacturing is the elimination of waste in all its forms. One of the ways to help ensure this happens is through Kaizen.

Kaizen is both the Chinese and Japanese word for continuous improvement (CI). When used in the business sense and applied to the manufacturing workplace, Kaizen refers to activities that continually improve all functions and involve all employees. By improving the standardization of all activities and processes, Kaizen aims to eliminate all wastes [10].

Kaizen is a cultural based strategy where all employees work together proactively to achieve regular, incremental improvements in the manufacturing processes. It combines the collective talents of a company to create an engine for continually eliminating waste from manufacturing processes. [4]

Kaizen or CI can be summarized by a four step iterative methodology for implementing improvements known as Plan, Do, Check, Act (PDCA) [4, 11].

PDCA defined in the context of an industrial manufacturing process improvement is:

1. Plan
    a. Establish objectives
    b. Set targets
    c. Determine expected results
2. Do
    a. Execute the plan
    b. Make the parts
    c. Collect data to use in the check and act steps
3. Check
    a. Chart the data and identify trends
    b. Analyze the results
    c. Determine differences between actual results and expected results
    d. Validate that the plan was correctly executed

4. Act
   a. Did the check show results better than expected?
      i. Yes implies that a new baseline has been found
      ii. This new baseline should be standardized
   b. Did the check show results worse than expected?
      i. Yes implies that there is more learning to be done
      ii. Planning for a follow-up PDCA event should be started
5. Repeat
   a. Repeat the PDCA process for ongoing continuous improvement

Figure 3-2 shows two trends for the Kaizen process. It is critical to standardize new baselines of improvement after each Kaizen event to ensure a lasting impact and faster overall improvement rate. Lack of standardization can lead to back-sliding and a longer, slower overall improvement rate [12].

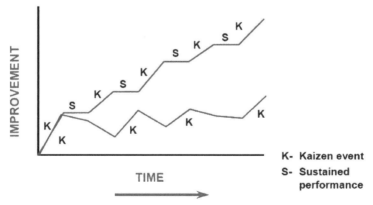

Figure 3-2. Sustaining the Kaizen Process

The "What is Lean?" Summary

Elimination of wastes is absolutely critical for running a lean manufacturing process. Wastes in manufacturing processes must be identified, must have visual alerts, must be measured and must show continuous improvement.

The four key areas discussed in this section will help with the following:
- Andon = Visual Alerts for Wastes
- Muda = Identify Wastes
- Overall Equipment Effectiveness (OEE) = Measure Wastes
- Kaizen = PDCA to continuously improve in the reduction of waste

WHY USE LEAN MANUFACTURING?

Questions to Consider about Lean

Four keys topics related to lean manufacturing and the elimination of waste were outlined in the previous section. Here are some questions to explore about lean manufacturing:

- Why would an energy consultant, an energy manager, a plant engineer, an industrial facility manager or a utility account executive with industrial customers have an interest in lean manufacturing?
- Why would your company or your customer want to invest time and money in lean manufacturing activities?
- What does lean manufacturing have to do with energy intensity?
- What is productive energy versus non-productive energy?

And finally, scrap:
- What's the big deal? We only scrap two or three parts a shift.

To answer these questions, we will take a look at some of the benefits of lean manufacturing and how this relates to energy consumption.

Why Have an Interest in Lean?

The first question relates to professionals in energy or plant engineering roles and their relationship to lean manufacturing.

Lean manufacturing by its very name appears to belong only on the shop floor, but involvement from energy and plant engineering teams is essential to lean success. Owners, stockholders, plant management, employees, state and local governments and utilities, all want to see long term growth for their industrial companies. They all want their industrial customers to:
- Be more competitive
- Thrive and add more jobs
- Grow and expand operations
- Stay where they are
- Make a good profit
- Pay dividends and pay taxes

By participating in the lean manufacturing process, energy and plant engineering professionals can help achieve these goals.

Why Invest in Lean?

Lean Manufacturing originated from the Toyota Production System (TPS). [13].

Typically, a lean manufacturing approach will provide the following:
- Better quality performance
- Fewer defects
- Less rework
- Less scrap
- Less waste
- Fewer machine breakdowns
- Fewer process errors
- Lower levels of inventory
- Greater levels of stock turnover
- More streamlined processes
- Better space utilization
- Higher process efficiencies
- Continuous improvement
- More output per man hour
- More output per energy unit input

- Increased capacity
- Improved delivery performance
- Faster changeovers
- Improved employee morale
- Better employee involvement
- Higher team commitment
- Improved supplier relations
- Lower operating costs
- Higher profits
- Better market share

Frequently the implementation of many lean manufacturing concepts is very inexpensive and can provide very significant savings. Lean manufacturing is a very good investment with excellent payback.

Profit is the difference between the selling price of an individual product and the cost to produce that product. The selling price is typically fixed by the market and largely out of your control. If you do not offer your product or service at the right selling price you will not retain or gain customers.

Cost is the area for focus and that is what lean manufacturing does. It focuses on elimination of wastes of all kinds. As costs go down, profits go up [13, 14].

On a macro view, profit is total sales minus total operating costs. Implementation of lean manufacturing principles can have a large impact on profit. Figure 3-3 is a graphic example of the potential impact of lean on profit. As operating costs go down, profits increase.

**What Does Lean Manufacturing
Have to do with Energy Intensity?**

Energy intensity is a measure of the amount of total input energy needed in terms of kilowatt-hours or Btus to create one unit of manufactured product. Examples of energy intensity are:
- Btu/ton
- kWh/linear yard

Lean Manufacturing Principles and Their Impact on Process Energy Efficiency

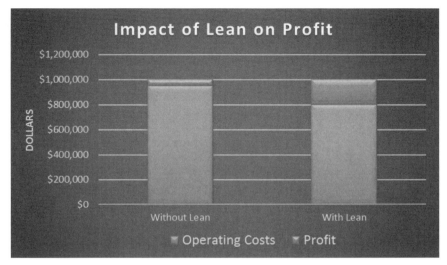

Figure 3-3. Example of the Impact of Lean on Profits

- kWh/pound
- Btu/part

Energy intensity can be used for tracking or trending a process to determine improvement. Also, it can be used for benchmarking a process for an indication of the relative energy use of one type of process versus another. For example, melting steel has a high process energy intensity compared to drying coatings on textiles.

The same lean tools that can measure and eliminate Muda (wastes) for a machine or a process, can also improve energy intensity for that same machine or process.

Energy intensity is a useful measure and can be incorporated with lean tools as a key process indicator.

What is Productive and Non-Productive Energy?

Lean manufacturing tries to eliminate all processes and actions that do not add value to the part.

In other words, if the process or action is not adding a value that the end customer is willing to pay for, then the process or action is a form of waste. Manufacturing processes and actions fall into two categories: value added and non-value added.

- For value added processes and actions, we want to:
  — Standardize
  — Optimize
  — Sustain
- For non-value added processes and actions, we want to:
  — Eliminate, if possible
  — Minimize, absolutely

Here is a quote about this concept:

"There is nothing quite so useless as doing with great efficiency something that should not be done at all." Peter Drucker [15]

Similarly, if energy is consumed by a process or action, but no value is added to the part that the customer is willing to pay for, then, again, this is waste. The usage of energy also has two categories: productive and non-productive energy. Here are some examples:

- A spindle servo motor turning a grinding wheel and engaging the surface of a part to make a final dimensional surface: productive energy

- That same spindle servo motor turning a grinding wheel all weekend long, with no part flow or part surface engagement: non-productive energy

- A motor on a hydraulic pump running at full load to turn the screw on a plastic injection molding machine to melt and inject the plastic into the mold: productive energy

- That same motor on a hydraulic pump, running and dead-heading during 24 of the 28 seconds of the injection molding machine cycle time when injection is not taking place: non-productive energy

- Compressing air up to 100 psi and then blowing it out of 1/8 inch holes in a metal pipe to blow the water off of gears coming out of a washer: non-productive energy
- The portion of the natural gas combustion heat content in a natural gas convection powder coat curing oven that goes directly up the exhaust stack at 1000°F with no heat recovery: non-productive energy

There are many, many more examples, but you get the idea. Just like the value and non-value added processes for lean manufacturing, there are productive and non-productive energy uses. These should be treated similarly:

- For productive energy usage, we want to
  — Optimize efficiency
  — Use the best available and most efficient technology
  — Consume energy only when adding value to the part
- For non-productive energy usage, we want to
  — Eliminate, if possible
  — Minimize, absolutely

Lean manufacturing tools can be equally useful for industrial processes and for optimizing energy usage.

Scrap: What's the Big Deal?

Have you ever heard someone in a manufacturing plant say "What's the big deal? We only scrap two or three parts a shift." Scrapping a part can be one of the most expensive and hardest to recover wastes in a manufacturing plant. This is especially true if the part is scrapped near the end of the production process after most or all of the value has been added.

At this point, you have already added loaded labor, possibly overtime labor, machine time, tooling, facility overhead, raw materials, and the energy used to add the value to the part. Now all of this is scrapped and all the value added is a financial loss and a waste. The worst part is that not only have you lost all the value in the scrapped part, but now, new material, labor, machine time, and energy must be expended to replace the scrapped part with another one.

Is two or three scrapped parts per shift a big deal? Absolutely, yes! Table 3-1 shows an example using a scrap rate of only two parts per shift.

Table 3-1. Impact of Scrap Cost

Item	Value	Explanation
Part	Fixed Outer Race (FOR)	A component for a front wheel drive automobile
Value Added	$12.00/FOR	Ready to go to assembly
Line Operations	1,050 shifts per year	Three shifts/day Seven days/week 50 weeks/year 3 x 7 x 50 =1,050
Shift Scrap Rate	Two FORs per shift	Production Data
Annual Scrap Rate	2,100 FORs per year	Two FORs/shift 1,050 shifts/year 2 x 1,050 = 2,100
Annual Scrap Cost	$25,200	2,100 FORs/year $12.00/FOR 2,100 x $12.00 = $25,200

This loss comes directly off the bottom line and takes away from company profits. Assuming a desired profit margin of five percent (5% or 0.05), the additional sales or cost cutting required to make up for this scrap loss is:

= $25,200/0.05
= $504,000 Yes, over half a million dollars!

Two or three parts per shift.... Yes, it is absolutely a big deal. Scrap is very bad. There is loss of value added, loss of materials, loss of labor, loss of transformation time and loss of energy. Additionally, as seen in the example above, scrap loss recovery is very expensive. Well beyond the cost of the scrap alone. Even a small amount of scrap over the course of a year can have a very dramatic impact on plant operating costs and profits.

The "Why Lean?" Summary

Lean manufacturing practices can help a company eliminate wastes, reduce costs and increase profits. It can be inexpensive and have rapid payback. Also, lean manufacturing principles can be applied to energy, as well as manufacturing activities and processes. The bottom line of lean manufacturing is that all waste can and must be eliminated.

So far we have learned about:
- Key lean manufacturing principles
- Why lean manufacturing is important and how it relates to energy
- The absolutely devastating cost of scrap

After looking at the "what" and the "why," let us now look at the "how." How can we make this lean impact happen? We will focus on the lean concept of Andon.

MORE ON ANDON

Focusing on Andon

The introduction to lean manufacturing principles and their importance has laid the groundwork for a deeper look into the concept of Andon. The primary purpose of this chapter is to provide information for energy professionals to use to make their industrial customers aware of the lean tool of Andon and its potential benefits to manufacturing processes. This chapter also provides information on specific Andon equipment and vendors. Andon, in its most basic form, is simply a visual signal. Andon (a Japanese term) literally means paper lantern [5]. Communicating by visual signals is very useful and is frequently used by the military. These include flashing light Morse code, semaphore and phonetic alphabet signal flags. One of the most common forms of an Andon system is a traffic light. An historic example of an Andon system was the two lanterns hung in the steeple of the old North Church in Boston to warn American troops that the British were invading by sea. Even Paul Revere understood the value of Andon.

Communicating with visual signals is a very useful lean manufacturing tool. Manufacturing plants can use Andon systems as a communication tool to display the status of a production line or process. Andon systems are found in many forms, from simple status lights to complete display boards in visible locations around the production plant. Andon systems can also include an audible alert system. They are a common tool in the automotive and component production industries. Andon is an integral part of the Toyota Manufacturing system [16]. Figure 3-4 shows a typical Andon color-coded stack light.

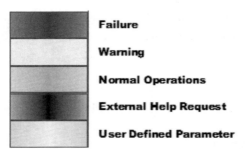

Figure 3-4. Typical Andon Color-coded Stack Light

What Does an Andon System Do?

Andon is an information tool which provides instant, visible and audible warnings to the manufacturing plant operations team that there is an abnormality occurring [17].

Andon is a technical installation supporting the execution of abnormality intervention by enabling industrial operation teams to:
- Better monitor equipment and processes
- Detect an abnormality
- Notify key troubleshooters of the abnormality
- Evaluate the severity of the abnormality
- Stop the equipment or process with the abnormality, if necessary

- Have an automated two-way communication system
- Fix or correct the abnormal condition, immediately
- Return the equipment or process to normal operation quickly
- Signal the return to normal operations with the "green" light
- Investigate the root cause and install a countermeasure [17, 18]

All of the above can lead to reduced Muda (wastes), improved OEE, and can guide priorities to perform Kaizen (PDCA/CI) on the applicable abnormalities.

The Andon system, by itself, does not:

- Solve abnormalities
- Prevent all defects from getting through to the next operation
- Replace good verbal communication between working groups
- Remove the need for corrective action and continuous improvement [17]

What Types of Companies Benefit from an Andon System?

Just about any industrial process line or individual piece of equipment could benefit from Andon.

Manufacturing plants with these specific operational characteristics can benefit even more from an Andon system:

- High energy cost (electricity, natural gas, fuel oil)
- High labor cost
- Low OEE
- High scrap rates
- High opportunity for defects
- High value added processes
- Continuous manufacturing
- High volume production
- Assembly line operations
- High value parts in a batch operation

Figure 3-5 shows the initiation of an Andon system process. There is a problem detection, the initiation of a visual Andon signal and then help arrives [19, 20].

What are the Features of an Andon System?

Andon systems can be customized to suit the particular manufacturing line or process that they are attached to. Here are some general features that most Andon systems have [16, 17, 19]:

- Machine operators can initiate a problem event at their work stations by pressing a button on the Andon terminal or pulling a cord.
- Some attributes can be programmed to automatically initiate an Andon event.
- Andon events can be activated by an error-proofing device.
- Visual and audible signals are automatically initiated, which can be seen and/or heard by the entire workgroup and workgroup leader.
- Andon systems have standardized light colors for the variety of situations (refer back to Figure 3-4).
- The Andon system automatically updates the stack light at the work station and the status light on the process or work cell Andon board.
- The Andon system sends out alerts about problem events using various messaging systems according to pre-defined notification rules and contact information.
  — Email
  — Text
  — Cell Phone or Radio Call
- Status can be monitored outside the production floor by using web-based interface.
- All events are recorded into a database for report generation, further processing, analysis and continuous improvement measures.
- Process and work cell data are updated in real time.

- Data can be displayed on flat screen monitors or light emitting diode (LED) displays.
- Allows for the escalation of a problem from green to yellow to red. Figure 3-5 shows an example of this.

What Information can an Andon System Provide?

One of the key features of an Andon system is that it can record data from equipment, process lines and events. These data are very useful for lean manufacturing activities and measurements. Some of this information can be provided directly and some is derived through calculations based on recorded data and operator input [16, 17, 18, 19]. Examples include:

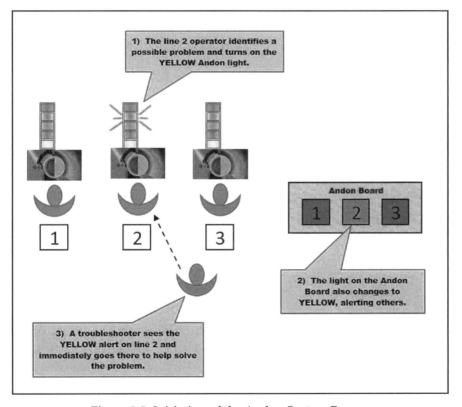

Figure 3-5. Initiation of the Andon System Process

- Real time actual production data versus production targets with selected unit of quantity (pieces, pounds, tons, linear yards, etc.)
- Totalized data to show actual production per hour, shift, day, week, month or year
- Machine and process operational parameters such as oil temperature, spindle speed, process temperature and pressure, feed rate, run time, energy consumption, etc.
- Auxiliary equipment setpoints such as chilled water temperature, compressed air system pressure, ambient air temperature and humidity, quench or coolant temperature, etc.
- OEE calculation from input on:
  — Availability
  — Performance
  — Quality
- Equipment downtimes with reason codes for sorting later
- Scrap and reason codes for sorting later
- Other key process indicators (KPIs)
- Overall plant production, quality, safety or delivery goals
- Type, duration, and number of problem events

The information from an Andon system is particularly valuable for selecting and prioritizing Kaizen events. An Andon system can reveal that the particular problem "X" occurs very often and causes significant downtime. Based on the Andon system data, proper resources can be allocated to solve the worst problems first.

Andon is not a substitute for leadership and management support. The hardware must be backed up by procedures, training, a culture of Kaizen, and operational team resources to respond to the Andon system and to analyze the Andon system data.

Setting up an Andon System

Hardware alone is only a very small part of a true company culture of Andon. Doing the proper planning and organizational homework is critical to the success of an Andon system. It is not productive to have an operator push the yellow Andon button, but have no one in the plant who knows what to do about it. This would be truly wasteful and inefficient, not to mention frustrating for the operator. An Andon system cannot stand alone. It is part of a much bigger culture of lean enterprise.

Here are some things to consider and implement before the first piece of Andon hardware is purchased [17, 18, 19]. Five Ws and One H for setting up an Andon system:

- WHO:
  — The Operations Team
- WHAT:
  — Determine what production machines and conditions need to be monitored
  — Determine the scope of the initial Andon system; one machine or the entire plant
  — Determine what type of Andon equipment will meet these needs
  — Determine which Andon signals are appropriate
  — Andon signals should be simple and easy for all to understand
- WHERE:
  — Divide the production system into logical areas
  — Assign each process or machine a number code and an owner
  — Assign small teams to each numbered process or machine
  — Post the number code and the owner/team names close to the process or machine
  — Determine the location of visual Andon boards
  — They should be visible from every corner of the relevant workplace

- WHY:
  — To facilitate the operator's ability to expose problems quickly
  — To facilitate the process of team problem solving to support the operators
  — To facilitate the collection of data for analysis and Kaizen

- WHEN:
  — After establishing escalation procedures
  — After identifying first responders and their roles and responsibilities
  — After ensuring proper training on and demonstration of Andon response
  — After establishing the process for analyzing Andon data for prioritizing Kaizen

- HOW:
  — Establish and implement the Andon procedures
  — Document the characteristics of the process and establish some historical data
  — Perform dry run walkthroughs of Andon response scenarios

Remember, Andon systems alone do not solve problems or reduce defects. People with training on team problem solving, Muda elimination, and Kaizen can use the data from an Andon system to solve problems and reduce defects. It is very important to prepare before making an investment in Andon equipment. Also, it may be worthwhile to just use a simple manual system of Andon to judge the value. Then it may be easier to justify the investment in a more elaborate system.

What are the Benefits of an Andon System?

Andon is a lean tool. Andon provides data that feeds other lean tools. Through the Kaizen process, Andon can have a positive impact on the key areas of reducing Muda, increasing OEE, and reducing energy intensity.

Following is a summary of both direct and somewhat intangible benefits of implementing an Andon system [16, 17, 18, 20]. With proper preparation, having an Andon system in place can:

- Give the operator an automated method to initiate problem resolution
- Allow reporting of problems immediately so that solutions can be quickly implemented
- Aid in quality control by having the ability to stop production and tackle quality issues as soon as they are identified
- Notify other workstation operators and supervisors of a problem
- Provide a complete picture of the overall status of the entire production line or cell
- Provide information to enable prioritization and coordination of corrective efforts
- Allow for problem escalation, as needed
- Facilitate line operator communications with team leaders and management
- Provide visual, real-time information
- Record data on:
  — Frequency and location of Andon activations
  — Reasons for Andon activations
  — Machine downtime and reasons
  — Scrap rates and reasons
  — Time to repair a failure
  — Mean time between failures
  — Other process specific items

By properly using the data available from the Andon system and the input of the operators involved, priorities can be set for conducting continuous improvement Kaizen events to tackle the

ANDON LIGHT	CONDITION	ACTION	WHO LEADS	HELP NAMES
RED	The process will *NOT* be completed on time!! We need MORE Help!!	Team Leader turns on the RED Andon light and contacts the PUM. The PUM coordinates help.	Production Unit Manager (PUM)	Engineering:_____ Logistics:_____ Materials:_____ Quality:_____
YELLOW	An issue has occurred that is threatening process completion. We need help!	Operator turns on the YELLOW Andon light. The Team Leader coordinates help.	Team Leader	Maintenance:_____ Other:_____
GREEN	The process is on schedule. There are no issues.	No Action	No Action	No Action

Figure 3-6. Escalation of an Andon Event

biggest problems first. Figure 3-6 shows a cycle of how the implementation of an Andon system can lead to continuous improvement.

Having an Andon system and properly using this data can have the following downstream benefits:
- Improved part quality
- Reduced scrap rate
- Elimination of all types of Muda
- Increased OEE with:
  — Better availability
  — Better performance
  — Better quality
- Identification of bottlenecks
- Reduction in work in process
- Reduction in overall inventory
- Better delivery and response time
- Avoid lack of parts
- Reduced energy intensity

All of these downstream benefits from the use of an Andon systems will also help to create:
- A culture of team problem solving and kaizen

- Increased production capacity
- Improve employee morale
- Reduced costs
- Improved profits

An Andon system has many benefits, some direct, some downstream. All of these benefits require a team of well trained and lean oriented employees. Again, preparations for Andon are crucial for success. Starting small and then expanding after proving the value may be a good approach. Andon systems do lend themselves to gradual implementation because of their:

- Simplicity
- Low Cost
- Scalability
- Quick Payback

Andon and Energy Professionals

There is definitely a direct correlation between the benefits of Andon and the goals of most energy professionals. Creating value is the common theme.

- If OEE is improved, then energy intensity is improved
- If less scrap is produced, then less non-productive energy is consumed
- If the plant quickly identifies and corrects problems on the line, then the better the overall process efficiency and the better the energy intensity
- The benefits from Andon and lean manufacturing also serve to improve energy efficiency and reduce energy intensity.

The energy professional can demonstrate his or her desire to help their customers by providing a simple overview of lean manufacturing and offering Andon as an idea to help improve processes. By working with an industrial customer on lean and

Andon topics, the energy professional can:

- Help the industrial customer to find financial incentives for energy efficiency projects.

- Improve the industrial customer's overall energy intensity

- Assist the industrial customer to incorporate lean and Andon into their Strategic Energy Management Planning (SEMP).

- Learn more about what the industrial customer does and what is important to them.

The Andon Summary

Andon is as much a philosophy as it is a system of hardware. Proper preparation and training is key to success with using Andon. Andon cannot solve problems by itself, but it is a great tool to collect data to prioritize continuous improvement Kaizen efforts.

So far we have learned about:

- Key lean manufacturing principles
- Why lean manufacturing is important and how it relates to energy
- The absolutely devastating cost of scrap
- Much more on the details and benefits of Andon

We will now focus on equipment and vendors for Andon systems.

ANDON EQUIPMENT AND VENDORS

Andon System Overview

Andon systems come in many forms and can range from very simple visual signals to complex interconnected computer driven systems. Figure 3-7 shows a basic Andon system operat-

Lean Manufacturing Principles and Their Impact on Process Energy Efficiency 105

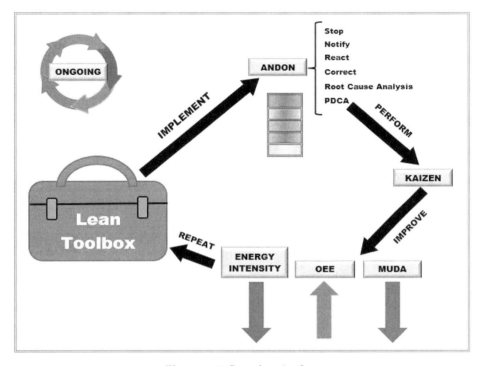

Figure 3-7. Ongoing Andon

The basic components of an Andon system include:

- A stack light on each monitored piece of equipment to indicate status
- A pull cord or push button for the operator to use to initiate and alert
- Speakers for audible alerts
- A local display panel for each line or work cell
- A wired or wireless network connected to a server
- Total plant display panels for overall plant status
- Mobile devices to receive alerts
- Radios to receive alerts

106 Industrial Energy Management Strategies

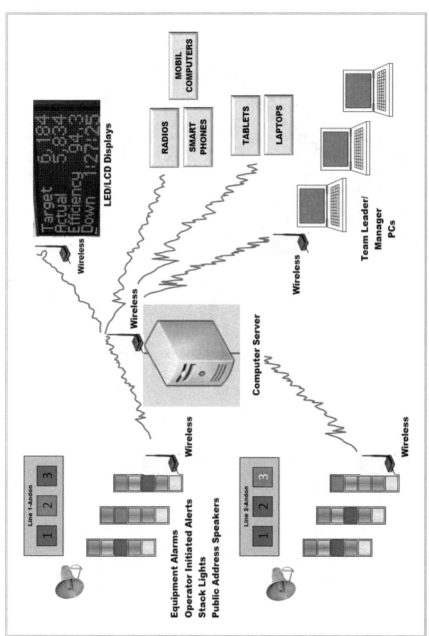

Figure 3-8. Basic Andon System Operating on a Wireless Network

- A computer server to interconnect devices and to store data
- Laptops and personal computers for observing and analyzing data

There are vendors that sell only some of the components for an Andon system. There are other vendors that sell turnkey systems with full system integration with the existing plant computer system. It is up to the individual manufacturing plant to decide the scope, size, connectivity, and complexity of their Andon system. In some cases, it may be wise to start with a small Andon system on one or two machines to test and evaluate the value. If this is successful, most systems are very easy to scale up.

The following sections contain specific information on vendors and their equipment, markets, locations and training resources. Including these vendors in this chapter is not an endorsement for their particular type of equipment. These vendors are listed here for informational purposes only and the following information was taken directly from the vendor websites. It is up to each manufacturing plant team to review and select the Andon vendor that is right for them.

Vendor 1-Actemium
Actemium Contact Information
      Actemium Process Automation North America
      6201 Fairview Road, Suite 200
      Charlotte, NC 28210
      Phone: (980) 215-8423
      Email: didier.micolaud@actemium.com
      Manager: Didier Micolaud
      Main website [21]: http://www.actemium.co.uk/en/ouroffers/mes/products/

Actemium Applicability to Andon
Andon is not the primary product, but it is a fully supported product line. There is a detailed file available:

- Actemium Solution Overview: An Introduction to Andon Systems [22]: http://www.actemium.co.uk/wp-content/uploads/2014/07/ActemiumAndonSolution.pdf

Actemium Company Information

Actemium headquarters are in the United Kingdom. With over 150 engineers and 30+ years of experience available to support Manufacturing Execution System (MES) and Supervisory Control and Data Acquisition (SCADA) requirements, Actemium is dedicated to delivering world class solutions to discrete manufacturing industries [21].

From automotive to assembly, aerospace to packaging, Actemium is one of only two Premier Solution Partners of GE Intelligent Platforms in the United Kingdom. Actemium can provide a complete MES solution from initial consulting to delivery and support of a solution.

Utilizing GE Intelligent Platforms' Proficy suite of industrial information technology software they can design, configure and implement a solution specifically tailored to their customers' requirements to deliver maximum business benefit from day one [21].

Actemium Andon Information

Manufacturers worldwide have implemented Andon solutions and achieved significant business benefits from doing so. The Andon systems not only support the concepts of visual factory and many of the soft benefits this provides, but also enable the collection and analysis of data to support on-going improvements in support of 6-Sigma and Lean Manufacturing initiatives [22].

Typically a solution of this type will help drive quality and efficiency improvements along with ensuring that all facilities are performing to the agreed specification in terms of cycle time, failure rate, etc. With customers who experience an average improvement in their plant efficiency the return on investment (ROI) of this solution is usually a simple payback of less than 12 months [22].

Actemium Market Segments
- Aeronautics, Space and Defense
- Automotive
- Chemicals
- Energy Generation
- Environment
- Feed, Food and Beverage
- Life Sciences
- Logistics
- Mining and Materials
- Nuclear Industry
- Oil and Gas
- Pulp and Paper
- Rail
- Refining
- Steel Processing [21]

Vendor 2-Entigral

Entigral Contact Information
    Entigral
    3716 National Drive, Suite 200
    Raleigh, North Carolina 27612 USA
    Phone: (919) 787-5885
    Toll Free: (877) 822-0200
    Fax: (919) 787-2522
    Email: info@entigral.com
    Main website: [23] http://www.entigral.com/

Entigral Applicability to Andon

    Andon equipment is not a product. The primary product is the TraxWare® software system for Radio Frequency Identification (RFID) entity tracking systems, which could be used in support of an Andon system installation.

- RFID is the wireless use of electromagnetic fields to transfer data, for the purposes of automatically identifying and tracking tags attached to objects. The tags contain electronically stored information [24].

Entigral Company Information

Entigral is headquartered in Raleigh, North Carolina adjacent to the Research Triangle Park. The company is expanding TraxWare as the Smart Manufacturing solution for the Internet of Things.

TraxWare efficiently tracks things in many complex environments and many workflow needs, most of which employ standard RFID tag and reader configurations. These "entity" tracking solutions became so relevant to the Internet of Things, and so "integral" to next generation enterprise processes, that the company changed its operating name to Entigral Systems, Inc. in 2008.

Entigral provides powerful and dynamic RFID and sensor management software

- RFID and next generation sensors for smart manufacturing
- Automate work-in-process solutions, intelligently
- Track inventory and goods throughout the manufacturing process
- Account for geographically distributed human and capital assets [23]

Entigral Andon Information

Today, Entigral focuses exclusively on the development, sale, and implementation of its TraxWare platform for asset, inventory, and work-in-process tracking in manufacturing and related sectors. It is not primarily related to Andon systems, but could be used in support of a plant-wide MES system with Andon characteristics.

Entigral Market Segments
- Manufacturing
- Government

- Power and Utilities [23]

Vendor 3-Leidos
Leidos Contact Information
    Charlotte Office
    15801 Brixham Hill Ave
    Suite 175
    Charlotte, NC 28277
    (704) 542-2908
    Raleigh Office
    615 Oberlin Road
    Suite 100
    Raleigh, NC 27605
    (919) 832-7242

There are many other locations worldwide.
Main website: [25] https://www.leidos.com/engineering/systems-integration-controls/andon

Leidos Applicability to Andon
    Andon is not the only or the primary product. Andon is an offering under the Systems and Controls group of the Leidos Engineering division [25]. There is one case study available, not directly applicable to Andon, but related to energy management:
- Case Study: Enterprise-Wide, Real-Time Energy Commissioning Services [26]. https://www.leidos.com/project/enterprise-wide-real-time-energy-commissioning-services

Leidos Company Information
    Leidos employees are inspired to create innovative technology solutions that solve the world's toughest problems in national security, health, and engineering. Through their culture of innovation, Leidos will develop deep customer trust and create enduring solutions that improve the world. Leidos is committed to investing in critical internal research and development efforts that respond to their customer's challenges, even as they work to

deliver the next generation of agile, cohesive solutions necessary for today's rapidly changing environment [27].

Leidos Andon Information

The Leidos Andon solution is a web-based production quality control and monitoring tool that supports assembly plant management. Functioning as a communications system, the Andon alert informs team members of the production status using emails, text messages, audio enhancements, and visual displays. This allows plant leaders and supervisors to quickly communicate issues, escalate problems to the appropriate personnel, and track historical performance to increase manufacturing uptime [25].

A good description of the Leidos Andon offering can be found here [28]: https://www.leidos.com/sites/default/files/Lcol-143_Andon_Systems.pdf

Leidos Market Segments
- National Security
- Health
- Engineering
  — Buildings and Facilities
  — Energy Management
  — Environmental
  — Oil, Gas and Chemical
  — Power Delivery
  — Transportation
  — Systems and Controls (Andon)
  — Utilities
  — Water and Waste [27]

Vendor 4-SCADAware

SCADAware Contact Information
　　　SCADAware
　　　1602 Rhodes Lane
　　　Bloomington, IL 61704
　　　Phone: (309) 665-0135

Toll Free: (888) 665-0135 (Support Desk)
Fax: (309) 665-0975
Email: info@scadaware.com
　　　sales@scadaware.com
　　　support@scadaware.com
Main Website: [29] http://www.scadaware.com/

SCADAware Applicability to Andon

Andon equipment and software is the primary product. Offerings include StatusWatch Production Monitoring Software, StatusLight Smart Andon, and call and response systems.

Training tutorials on the StatusWatch Production Monitoring Software are available. StatusWatch Video Tutorials [30]: http://www.scadaware.com/statuswatch-video-tutorials/

The SCADA in this vendor's name SCADAware is a common computer system acronym and stands for: Supervisory Control and Data Acquisition. SCADA is a system operating with coded signals over communication channels so as to provide control of remote equipment typically using one communication channel per remote station [31].

SCADAware Company Information

SCADAware is a systems integration firm comprised of a uniquely innovative and talented team of engineers and programmers who specialize in the design and implementation of open-architecture control systems and PC/PLC based SCADA systems for heavy industry, investor-owned utilities, municipal water/wastewater utilities, and government agencies.

After decades of hands-on experience in plant operations, StatusWatch Product Suite was developed to monitor asset utilization and analyze plant efficiency. Companies around the world rely on this product to make more money. Through intelligent production monitoring, timely decision making, improved accountability and OEE calculations, SCADAware products help companies optimize their flow and efficiency by decreasing waste and increasing value [29].

SCADAware Andon Information

Figure 3-9 SCADAware StatusWatch Product Suite is an excerpt from the SCADAware main website [29] and show some of their key Andon products.

StatusWatch Product Suite

StatusWatch Production Monitoring Software
- Delivers timely, accurate, and detailed information on asset usage and plant efficiency.
- Unlimited custom reports and flexible notifications allow users to enhance output and produce top quality products.
- Charts can represent both real-time and historical data critical to the lean manufacturing process.

StatusLight Smart Andon
- Highly visible stack lights allow users to react faster to actionable events.
- A point-and-click graphical interface makes it easy to configure everything from complex light relations to barcode inputs.
- Fully customizable light behaviors allow users to assign unique significances to user input.
- StatusLights integrate seamlessly with StatusWatch.

StatusLight Call and Response System (CRS) Application
- Allows computers to receive materials and maintenance requests directly from StatusLights.
- Requests can be viewed, responded to, and signaled complete through the CRS application.
- All CRS activities are tracked and time-stamped in StatusWatch.

StatusWatch Line Sign Application
- Allows users to see more of their facility at a glance.
- The Line Sign displays the status of many machines or andon devices in real-time.
- View any number of screens in rotation and know instantly where help is needed.

Figure 3-9. Scadaware Statuswatch Product Suite

SCADAware Market Segments
SCADAware has proven experience in the following industries:
- Water and Wastewater Treatment
- Manufacturing
- Gas Transmission Pipelines
- Military/Government
- Food and Beverage
- Facility Automation
- Wood Treating
- Logistics and Distribution
- Aerospace [29]

Vendor 5-SeQent

SeQent Contact Information
    SeQent Corporate Office
    4500 Blakie Road, Suite 137
    London, Ontario
    N6L 1G5
    Canada
    Phone: (519) 652-0401
    Fax: (519) 652-9275
    Email: information@SeQent.com
    Main website: [32] http://www.seqent.com/index.php

SeQent Applicability to Andon

Andon related equipment is a main product and is primarily related to plant floor messaging and communications systems. There are two related case studies:

- Automotive Tier 1 Supplier Case Study

  Wireless Alarm & Event Dispatch Reduces Downtime and Asset Reliability [33] http://www.seqent.com/PDFs/Case%20StudyAutomotive.pdf

- Andon-Visual Display Management-Bottling Case Study [34] http://www.seqent.com/PDFs/Bottling.pdf

SeQent Company Information

For a brief overview of what SeQent products can do, there is a video on the website home page: Andon-Industrial-Digital-Signage Solutions [32] http://www.seqent.com/index.php.

SeQent is privately held and is headquartered in London, Ontario, Canada, with offices in the United States and is the leading provider of plant floor messaging and communications equipment. Leveraging its Real-Time Decision Solutions (RTDS) platform—SeQent provides a hardware/software bridge from alarms and events that occur in plant floor equipment and applications and sends real-time build counts, faults, targets, actuals, OEE and KPIs to textual, visual and audible outputs.

SeQent has become the recognized expert in the field of wireless alarm and event dispatch, Andon-Industrial Digital Signage and Wireless KanBan solutions with partnerships and technology exchanges with industry giants such as Cisco, Motorola, Activplant, GE Intelligent Platforms and Rockwell Automation. SeQent customers include some of the largest discrete and process manufacturers in the world including:

- Abbott Nutrition
- Chrysler
- Ford Motor Company
- General Motors
- Georgia Pacific
- Honda
- Intel
- Kellogg's
- NBC/Universal Studios
- Proctor and Gamble
- SC Johnson
- Toyota
- WS Packaging Group [32]

SeQent Andon Information

SeQent's solutions help many of the world's largest manufacturers produce their products more efficiently and to high quality standards. SeQent's product lines are sold through an extensive array of OEM private label and reseller agreements, partnerships with automation supplier sales channels and directly to end users. SeQent's solutions drive continuous improvement initiatives by delivering the following benefits:

- Cost reduction
- Increased productivity
- Improved quality
- Reduced staffing requirements
- Employee motivation

- Improved IT utilization
- Reliability
- Operational visibility
- Lower inventories
- Quick ROI
- Reduced TAKT time [32]

Figure 3-10 shows how SeQent connects all types of Andon and other devices [35].

SeQent Market Segments
- Automotive/Aerospace Manufacturers
- Automotive/Aerospace Suppliers
- Consumer Packaged Goods
- Food and Beverages [32]

Vendor 6-VersaCall
VersaCall Contact Information
    VersaCall Technologies Inc.
    7047 Carroll Road
    San Diego, Ca 92121
    Phone: (858) 677-6766
    Fax: (858) 677-6765
    Email: sales@versacall.com

VersaCall Sales and General Support
    Phone: (858) 677-6766 (option 1)
    Email: sales@versacall.com

VersaCall Technical Support (Hardware)
    Phone: (858) 677-6766 (option 2)
    Email: support@versacall.com

VersaCall Technical Support (Software)
    Phone: (858) 677-6766 (option 3)
    Email: support@versacall.com
    Main website: [36] http://versacall.com/

118   Industrial Energy Management Strategies

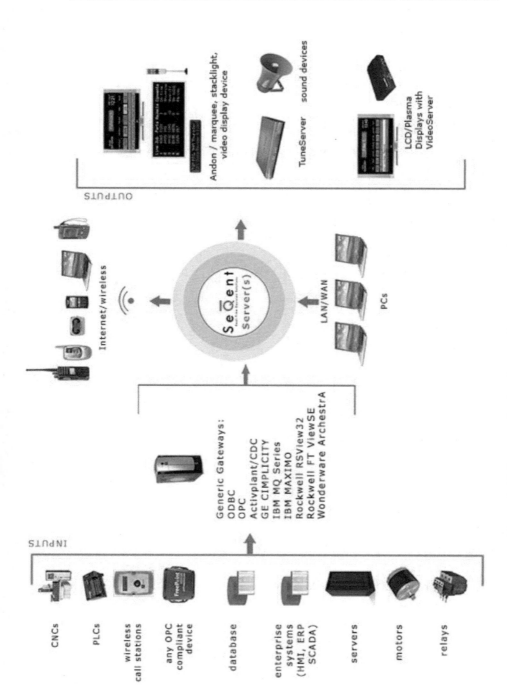

Figure 3-10. Sequent Andon System Components and Arrangement

VersaCall Applicability to Andon

Andon related equipment is the main product line including hardware and software for Andon wireless systems, call systems, visual management and OEE systems. There are five related case studies, all at the same link [37].

Link to the five case studies: http://versacall.com/case-studys-success-stories/
- 3M-Recovery of Lost Production
- CAT-Immediate Communications
- Hamilton Sunstrand-Visibility of Assembly Downtime Issues
- Honeywell-Turning Downtime into Productive Time
- Harley-Davidson-Audio and Visual Andon Systems

VersaCall Company Information

Since 1995, VersaCall Technologies Inc. has a legacy of being the leader in developing wireless industrial systems and solutions that improve productivity and operations efficiency:
- Formerly the Facility Communications Group of Indyme Solutions
- Continually developing innovative wireless solutions for reducing production cost and improving production output
- Systems engineered to withstand the demands of the factory floor environment
- Each system is tailored to a site's specific requirements
- Hardware and software modules are designed to be configurable, flexible and expandable

VersaCall Wireless Systems have been effective in reducing downtime on the production floor by as much as 35 percent. Their wireless technology is easy to implement and integrates well with existing or planned manufacturing processes. One hundred percent ROI in as little as 8-16 weeks is possible. Providing immediate value is the hallmark of a VersaCall System [36].

VersaCall Andon Information

Manufacturing plant monitoring and communicating is VersaCall's bread and butter. They offer a wide range of hardware components and software systems to support Andon related implementation.

Hardware offerings include [36]:
- Core System Modules
  — Standard Control Unit
  — Mini-Control Unit
  — Receiver/Coordinator
  — Repeater

- Input Modules
  — Call Stations
  — Reason Code Modules
  — Data Input Modules
  — Assembly Modules

- Output Modules
  — Integrated Stack Lights
  — Wireless Stack Lights
  — Wireless Audio Modules
  — Wireless Control Modules

- Communication Modules
  — Two-Way Radio Modules
  — On-Site Paging Systems
  — Ethernet Interfaces

Software offerings include [36]:
- Display Software-Virtual Panel III Software
  — Supports the display of performance information on:
    - Smart Phones
    - Tablets
    - Desktop Computer Displays
    - Large Screen Displays
  — Supports the development of:
    - Messaging Panels
    - Performance Panels
    - Communication Panels

- Reporting Software-VersaCall
  — Supports the development of:
    - Automated Reports
    - Preconfigured Reports
    - AdHoc Reports
- OEE Software
  — Supports the development of:
    - Standard OEE Reports
    - Custom OEE Reports
- Custom and Special Application Software Development

For a brief overview of some product capabilities, there are two videos on the website home page:
- The VersaCall VT3000 System [38] http://versacall.com/
- The Innovative Wireless System [39] http://versacall.com/

For an overview of a six-step process to get started with an Andon type system, there is a process at this link: Getting Started-Six Steps to Developing an Application for Your Site [40] http://VersaCall.com/6-easy-steps/

VersaCall Market Segments
A broad range of industrial sites, including:
- Foundries
- Food processors
- Truck assembly plants
- Medical products
- Office equipment
- Motorcycles
- Heavy equipment [36]

Vendor 7-Vorne
Vorne Contact Information
Vorne Industries, Inc.
1445 Industrial Drive
Itasca, IL 60143

Phone: (630) 875-3600
Toll Free: (877) 767-5326
Fax: (630) 875-3609
Email: Sales: sales@vorne.com
Support: support@vorne.com
Main website [41]: http://www.vorne.com/

Vorne Applicability to Andon

Andon related equipment is the main product with a focus on data collection and display hardware [41]. Vorne has an extensive learning center on their website with detailed information on lean manufacturing principles including Andon, Kaizen, OEE, The Visual Factory and others. These detailed learning topics can be found at this link:

Vorne Learning Center [42] http://www.vorne.com/lean-learning-center.htm

Vorne Company Information

Vorne Industries Inc., works hard to bring their customers the very best productivity enhancing products available, period. They listen to their customers and respond with well-engineered solutions to real world problems. Vorne is not a distributor, they manufacture their own innovative product designs and maintain complete control over design, quality, lead-time and support.

Since 1970, Vorne's passion has been to provide productivity enhancing products that work harder, last longer and provide better value than anything else available. They know what works and what doesn't for most any industry and are happy to help.

Vorne data collection and visual display products are used around the world at thousands of companies representing nearly every industry. They use application experience and ongoing lean manufacturing research to provide above and beyond value at a great price [41].

Vorne Andon Information

Vorne focuses on improving manufacturing plant operations. They have years of plant experience and build key lean manufacturing capabilities into their equipment. Their solutions focus on the following actions:

- Increase Capacity
- Measure OEE
- Reduce Changeover Times
- Reduce Cycle Times
- Reduce Down Time
- Reduce Manufacturing Costs
- Improve Productivity
- Display OEE
- Display Production Counts
- Display KPIs on Scoreboards [41]

One of their key Andon related products is the XL Productivity Appliance™ [43].

The XL is an extremely flexible and powerful tool for improving manufacturing productivity. XL is a "bolt-on" smart device that provides you with plant-wide real-time data visibility and a comprehensive set of performance management tools. It combines six products in one simple package including a visual display, production monitor, data warehouse, and input/output processor, embedded server and programming platform.

Figure 3-11 shows an example of one Vorne's products [44].

Vorne Market Segments
Vorne products are in literally thousands of manufacturing plants across all types of industries [41].

Vendor 8-Werma
Werma Contact Information
    WERMA USA Inc.
    6731 Collamer Road
    East Syracuse, NY 13057 USA

Figure 3-11. Example of a Vorne Display Board

Phone: (315) 414-0200
Fax: (315) 414-0201
E-Mail: us-info@werma.com
Main website [45]: http://www.werma.com/en/

Werma Applicability to Andon

Andon related equipment is the main focus including stack lights, optical signal devices, audible signal devices and lean management system components [45]. Werma has a company newsletter called the Werma Report. The issue number 25 from July 2013 has an article related to Andon. Production Efficiency with Werma Signaltechnik:

"New Andon products and the WIN™ system as integral tools for the implementation of lean production methods" [46]
http://www.werma.com/gfx/file/report/2013_WERMA-Report-Nr25_EN_Web.pdf

The WIN™ system acronym used by Werma stands for "Wireless Information Network" [47]

Werma Company Information

WERMA Signaltechnik is one of the world's leading companies in the design and manufacture of equipment commonly known as beacons and sounders. The global company, based in southwest Germany, sets the pace with its many innovative design solutions. Werma signal devices ensure that the working environment is safe and processes efficient, whether on machines or within process systems, in the factory, in the office, or in public spaces. From signal towers, beacons and warning lights to sounders, buzzers and horns, Werma optical, audible and explosion proof signal devices warn and protect people throughout the world.

Werma has been developing and manufacturing electrical devices for over 60 years. Their signal devices inform and alert people, machines and computers. With a wide range of over 3,500 items, signal devices from Werma offer solutions for the most varied of applications. Werma has had a location in the USA as of 2012.

Werma Andon Information

Detailed information on Werma capabilities and equipment related to Andon and lean manufacturing can be found in this brochure [47]:

Improve Operation Efficiency Lean Production Solutions: http://www.werma.com/gfx/file/report/2013_Lean-Production-Loesungen_EN_int.pdf

Werma Market Segments

Most any industry segment, production line, or industrial process could use Werma products [47].

Vendor Summary Tables

There are two tables included here to allow for cross referencing between vendors, market segments and Andon related applications:

- Table 3-2 shows market segments versus applicable Andon vendors.

- Table 3-3 shows Andon applications versus applicable Andon vendors.

Table 3-2a. Market Segments Versus Applicable Andon Vendors

Market Segments	Applicable Andon Vendors
• Aerospace	Actemium Entigral Leidos SCADAware SeQent VersaCall Vorne Werma
• Automotive	Actemium Entigral Leidos SCADAware SeQent VersaCall Vorne Werma
• Chemicals	Actemium Entigral Leidos Vorne Werma
• Control Systems	Actemium Leidos SCADAware SeQent
• Food and Beverage	Actemium Entigral SCADAware SeQent VersaCall Vorne Werma

Table 3-2b. Market Segments Versus Applicable Andon Vendors

Market Segment	Vendors
• Mining and Minerals	Actemium Vorne Werma
• Oil and Gas	Actemium Leidos
• Packaging	Entigral SeQent Vorne Werma
• Primary Metals	Actemium VersaCall Vorne Werma
• Pulp and Paper	Actemium Entigral Vorne Werma
• Refining	Actemium Leidos Vorne Werma
• Transportation (not Automotive) 　○ Heavy Equipment 　○ Trucks 　○ Motorcycles 　○ Rail	Actemium Entigral Leidos VersaCall Vorne Werma
• Buildings and Facilities	Leidos SCADAware Werma

Table 3-2c. Market Segments Versus Applicable Andon Vendors

Market Segment	Vendors
• Government, NASA, Military and National Security	Actemium Entigral Leidos SCADAware
• Logistics and Distribution	Actemium SCADAware Vorne Werma
• Utilities, Power Generation and Delivery, Energy Management, Nuclear Industry and Gas Pipelines	Actemium Entigral Leidos SCADAware
• Water and Wastewater, Environmental, Health and Life Sciences	Actemium Leidos SCADAware Vorne Werma

SUMMARY AND CONCLUSIONS

Target Audience
- Energy professionals
- Industrial manufacturing plant personnel

Purpose
- To provide information for energy professionals to use for the benefit of their industrial customers through education and implementation of lean manufacturing principles, particularly Andon systems.
- To help energy professionals assist their industrial customers to understand lean manufacturing concepts and the associated energy and non-energy benefits.

Table 3-3. Andon Applications Versus Applicable Andon Vendors

Andon Applications	Applicable Andon Vendors
Turnkey Andon System Solutions	Actemium Leidos SCADAware VersaCall Vorne Werma
Andon and Related Software	Entigral SCADAware VersaCall
Andon System Components • Stack Lights • Audible Alerts • Modules • Displays, etc.	SCADAware SeQent VersaCall Vorne Werma
Andon Connectivity • Wireless • Messaging • Communications • Servers, etc.	SCADAware SeQent VersaCall Vorne Werma

Topics
- Introduce four key lean manufacturing topics
  — Andon
  — Muda
  — OEE
  — Kaizen
- Provide reasons for implementing lean manufacturing
- Relate lean manufacturing to energy savings
- Demonstrate the severe financial impact of scrap
- Provide details on the Andon lean manufacturing principle
- Provide details on selected vendors for Andon equipment and systems

Key Points to Remember
- The benefits of lean manufacturing principles are proven and well documented.
- Basically, lean principles focus on:
  — Elimination of waste
  — Reduction of costs
  — Improvement in profits
- Lean principles have the same effect on energy usage:
  — Elimination of non-productive energy (waste)
  — Reduction of energy costs
  — Reduction in energy intensity
- Andon is a key principle for lean manufacturing.
- Andon provides a way to see, react to and document manufacturing problems.
- Andon provides data for continuous improvement processes
- Andon implementation can be:
  — Inexpensive
  — Quick to payback
  — Simple
  — Started small
  — Scaled up easily
- There are lots of Andon vendors that can provide components up to a turnkey system
- Electrical utilities should be aware of and interested in lean and Andon systems for the benefit of their industrial customers.

GLOSSARY

Glossary Context

All terms in this glossary are defined in the context of lean manufacturing activities in industrial plants. These terms may have other meanings outside the context of the manufacturing environment.

Glossary Terms

Andon [2, 3, 5]:
1. The Japanese word for paper lantern.
2. A manufacturing term referring to a system to notify team leaders, managers, maintenance technicians or other workers of a quality, material or process problem.
3. A visual system that incorporates color coded lights and real time communication alerts and displays, which initiate immediate help for the emerging problem.
4. A signal, light, bell or music alarm, triggered by an operator confronted with a non-standard condition.

Andon Board [2, 3, 5]:
1. A visual control device in a production area giving the current status of the production processes or equipment.
2. A visual alert system to team members of emerging problems.
3. The information is often displayed in the colors green, yellow and red.

Continuous Improvement [2, 3, 4, 11]:
1. A never-ending effort to expose and eliminate root causes of problems.
2. A four-step iterative methodology for implementing improvements known as Plan, Do, Check, Act (PDCA).
3. Synonym: Kaizen.

Energy Efficiency:
1. The ratio of useful energy output divided by the total energy input for a system as a percentage less than 100%.
2. The (total energy input – losses)/total energy input, also as a percentage less than 100%.

Energy Intensity:
1. A measure of the amount of energy needed in terms of kilowatt-hours or Btus to create one logical unit of manufac-

tured product, e.g. Btu/ton, kWh/linear yard, kWh/pound.
2. An indicator of the relative energy use of a process or industry when compared to other energy intensities, e.g. melting steel has a high process energy intensity compared to drying coatings on textiles.

Kaizen [2, 3, 4, 10]:
1. Both the Chinese and Japanese word for continuous improvement.
2. A cultural based strategy where all employees work together proactively to achieve incremental, regular, continuous and "do it now" improvements in manufacturing processes.
3. Activities that continually improve all functions, involve all employees and eliminate all Muda (wastes).
4. Synonym: Continuous Improvement.

Kaizen Event [2, 3, 4]:
1. A concentrated effort, in which a team plans and implements a major process change to quickly achieve a quantum improvement in performance.
2. Participants generally represent various functions and perspectives, and may include non-plant personnel.

KanBan [2, 4]:
1. A communication tool in the "just-in-time" production and inventory control system which authorizes production or movement of materials.
2. A "pull production" means of communicating the need for products or services.

Lean Manufacturing [1, 2, 3]:
1. A production philosophy that considers the expenditure of resources for anything other than the direct creation of value for the end customer to be wasteful, and therefore a target for elimination.
2. The trimming away of all the Muda (wastes), which leaves

in place an optimized process made up of value adding essential actions only.

Muda [2, 3, 4, 6]:
1. The Japanese word meaning uselessness, idleness, or literally waste.
2. Activities that consume resources, but add no value.
3. Commonly categorized in lean manufacturing practices into seven areas, plus one:
   a. Seven Areas:
      i. Transportation
      ii. Inventory
      iii. Motion
      iv. Waiting
      v. Over-Processing
      vi. Over-Production
      vii. Defects
   b. Plus one:
      i. Ineffective use of employee talent
4. Synonym: Waste

Non-Productive Energy:
1. A term similar in meaning to non-value added, but related to energy inputs and wastes.
2. The energy that is consumed by an activity that does not add value to the product.
3. Non-productive energy should be eliminated, if possible, and otherwise minimized.
4. Antonym: Productive Energy.

Non-Value Added [2, 3, 4]:
1. Activities which are essential tasks that have to be done under present working conditions, but do not add value to the product (sometimes referred to as required waste).
2. Activities that the end customer will not and does not pay for.

3. The desire is to either minimize these activities or introduce process improvements that would eliminate them entirely.
4. Antonym: Value Added.

Overall Equipment Effectiveness (OEE) [2, 3, 7, 8, 9]:
1. A methodology to measure the performance of a specific machine or a production line.
2. OEE is made up of the mathematical product of three percentage factors:
   a. Availability = {Actual Operating Time/Scheduled Operating Time} x 100
   b. Performance = {(Parts Produced x Ideal Cycle Time)/Actual Operating Time} x 100
   c. Quality = {(Parts Produced – Defective Parts)/Parts Produced} x 100
3. A common target level for world class OEE is 85 percent.

Productive Energy:
1. A term similar in meaning to value added, but related to energy inputs and wastes.
2. The energy that is consumed by an activity that does add value to the product.
3. To maximize productive energy:
   a. Optimize equipment/process efficiency
   b. Use the best available and most efficient technology
   c. Consume energy only when adding value to the part
4. Antonym: Non-Productive Energy.

Value Added [2, 3, 4]:
1. A type of processing (accomplished correctly the first time) that
   a. Transforms the fit, form, or function of a product or assembly, and
   b. That the end customer is willing to pay for.
2. If the process or action is not adding a value that the end customer is willing to pay for, then the process or action is a

form of Muda (waste), and must be eliminate or minimized
3. Value added processes and actions must be:
   a. Standardized
   b. Optimized
   c. Sustained
4. Antonym: Non-value Added.

Bibliography

1. Wikipedia. (2015, May 21). Lean Manufacturing. Retrieved from Wikipedia: The Free Encylopedia: https://en.wikipedia.org/wiki/Lean_manufacturing
2. Gembutsu Consulting. (2009). 7. Lean Manufacturing Glossary, Definitions and Terms. Retrieved from Gembutsu Consulting LLC: http://www.gembutsu.com/articles/leanmanufacturingglossary.html#g
3. Lake Region Manufacturing. (n.d.). Lean Manufacturing Definitions Made Easy. Retrieved from Lean Sigma Definitions: http://www.lakeregionmedical.com/transition/pdfs/LeanSigmaDefinitions.pdf
4. Vorne Industirs Inc. (2013). Top 25 Lean Tools. Retrieved from Lean Production: http://www.leanproduction.com/top-25-lean-tools.html
5. Wikipedia. (2015, June 4). Andon (manufacturing). Retrieved from WIkipedia: The Free Encyclopedia: http://en.wikipedia.org/wiki/Andon_%28manufacturing%29
6. Wikipedia. (2015, May 27). Muda. Retrieved from Wikipedia: The Free Encyclopedia: https://en.wikipedia.org/wiki/Muda_%28Japanese_term%29
7. Howard, William S. (2009). Overall Equipment Effectiveness. Retrieved from Stability Technology: http://www.stabilitytech.com/lean_measure.html
8. Wikipedia. (2015, June 1). Overall Equipment Effectiveness. Retrieved from Wikipedia: The Free Encyclopedia: https://en.wikipedia.org/wiki/Overall_equipment_effectiveness
9. OptimumFX Consulting. (n.d.). Overall Equipment Effectiveness-OEE Explained. Retrieved from OptimumFX Consulting: http://www.optimumfx.com/overall-equipment-effectiveness-oee-explained/
10. Wikipedia. (2015, June 3). Kaizen. Retrieved from Wikipedia: The Free Encyclopedia: https://en.wikipedia.org/wiki/Kaizen
11. Wikipedia. (2015, May 28). PDCA. Retrieved from Wikipedia: The Free Encyclopedia: https://en.wikipedia.org/wiki/PDCA
12. Armendez, L. (2009). Work Measurement and Lean Applications in the Process Industries. Retrieved from Institute of Industrial Engineers: https://www.iienet2.org/Details.aspx?id=17506
13. Earley, T. (2015). Benefits of Lean Manufacturing | Why Implement Lean? Retrieved from Lean Manufacturing Tools: http://leanmanufacturingtools.org/63/benefits-of-lean-manufacturing/
14. McKewen, Ellen. (2012, August 7). Top 4 Reasons for Implementing Lean Manufacturing. Retrieved from CMTC Manufacturing Blog: http://www.cmtc.com/blog/bid/132583/Top-4-Reasons-for-Implementing-Lean-Manufacturing
15. Shore, J. (2014, September 16). These 10 Peter Drucker Quotes May Change Your World. Retrieved from Entrepreneur: http://www.entrepreneur.com/article/237484
16. leanmanufacture.net. (2009). Andon Systems in manufacturing and process op-

erations. Retrieved from lean manufacturing & operating management: http://www.leanmanufacture.net/leanterms/andon.aspx
17. Beyond Lean. (n.d.). Module: Autonomation Element: Andon Systems. Retrieved from Beyond Lean: www.beyondlean.com/support-files/andon.pdf
18. Andon Production Line. (2011). Andon Industrial Production Line Monitoring System. Retrieved from Andon Production Line: http://www.andon.pl/index.html
19. Velaction Continuous Improvement. (2014). Andon (+7-min MP#, +6-Page PDF). Retrieved from Velaction Continuous Improvement: http://www.velaction.com/lean-andon/
20. Velaction Continuous Improvement. (2014). Andon Process Summary (Infographic). Retrieved from Velaction Continuous Improvement: http://www.velaction.com/andon-process-summary-infographic/
21. Actemium. (2015). MES-Manufacturing Execution Systems. Retrieved from MES Products: http://www.actemium.co.uk/en/ouroffers/mes/products/
22. Actemium. (2014, February 12). An Introduction to Andon Systems. Retrieved from Actemium Solution Overview: http://www.actemium.co.uk/wp-content/uploads/2014/07/ActemiumAndonSolution.pdf
23. Entigral. (2015). Innovative Solutions for the Smart Manufacturing Future. Retrieved from Traxware Connects Things: http://www.entigral.com/
24. Wikipedia. (2015, June 11). Radioi-frequency identification. Retrieved from Wikipedia: The Free Encyclopedia: https://en.wikipedia.org/wiki/Radio-frequency_identification
25. Leidos. (2015). Andon Systems. Retrieved from Leidos Engineering: https://www.leidos.com/engineering/systems-integration-controls/andon
26. Leidos. (2015). Enterprise-Wide, Real-Time Energy Commissioning Services. Retrieved from Leidos Engineering: Case Study: https://www.leidos.com/project/enterprise-wide-real-time-energy-commissioning-services
27. Leidos. (2015). A New Perspective. Retrieved from Corporate Capabilities Brochure: https://www.leidos.com/sites/default/files/Corporate%20Capabilities%20Brochure-Lite-A.pdf
28. Leidos. (n.d.). Andon System PDF. Retrieved from Leidos Lcol-143: https://www.leidos.com/sites/default/files/Lcol-143_Andon_Systems.pdf
29. SCADAware. (2015). SCADAware. Retrieved from Transforming Automation for Continuouos Improvement: http://www.scadaware.com/
30. SCADAware. (2015). StatusWatch. Retrieved from StatusWatch Video Tutorials: http://www.scadaware.com/statuswatch-video-tutorials/
31. Wikipedia. (2015, May 24). SCADA. Retrieved from Wikipedia: The Free Encyclopedia: https://en.wikipedia.org/wiki/SCADA
32. SeQent. (2014). Real-Time Decision Solutions. Retrieved from SeQent: http://www.seqent.com/index.php
33. SeQent. (n.d.). Automotive Tier 1 Supplier Case Study. Retrieved from SeQent Case Study: http://www.seqent.com/PDFs/Case%20StudyAutomotive.pdf
34. SeQent. (n.d.). Andon-Visual Display Management-Bottling Case Study. Retrieved from SeQent Case Study: http://www.seqent.com/PDFs/Bottling.pdf
35. SeQent. (n.d.). Andon System Block Diagram. Retrieved from SeQent-big-chart: http://www.seqent.com/our-solutions/SQ-big-chart.php
36. VersaCall. (2015). Wireless Innovations Improving Productivity. Retrieved from VersaCall Home Page: http://versacall.com/
37. VerasCall. (2015). Case Study's/Success Stories. Retrieved from VersaCall Case Studies: http://versacall.com/case-studys-success-stories/

38. VersaCall. (2015). VT3000 Video. Retrieved from VersaCall Home Page: http://versacall.com/
39. VersaCall. (2015). Innovative Wireless Systems Video. Retrieved from VersaCall Home Page: http://versacall.com/
40. VersaCall. (2015). Six Steps to Developing an Application for Your Site. Retrieved from VersaCall-Getting Started: http://versacall.com/6-easy-steps/
41. Vorne. (2015). Let the Improvements Begin. Retrieved from Vorne Homepage: http://www.vorne.com/index.html
42. Vorne. (2015). Vorne Learning Center. Retrieved from Vorne Learning Center: http://www.vorne.com/lean-learning-center.htm
43. Vorne. (2015). XL Productivity Appliance™. Retrieved from Vorne XL: http://www.vorne.com/xl/index.html
44. Vorne. (2015). OEE Displays in Manufacturing. Retrieved from Vorne OEE: http://www.vorne.com/solutions/oee-displays.htm
45. Werma. (2015). Werma Signaltechnik. Retrieved from Werma Home Page: http://www.werma.com/en/
46. Werma. (2013, July). Production Efficiency with Werma Signaltechnik. Retrieved from Werma Report, No. 25: http://www.werma.com/gfx/file/report/2013_WERMA-Report-Nr25_EN_Web.pdf
47. Werma. (2013). Improve Operation Efficiency: Lean Production Solutions. Retrieved from Werma Brochure: http://www.werma.com/gfx/file/report/2013_Lean-Production-Loesungen_EN_int.pdf

Chapter 4

Operational Savings

One of the biggest benefits of implementing a sustained energy management program is the operational savings. Studies from the U.S. Department of Energy's Superior Energy Performance program support the implementation of best practices that show an average energy reduction between 5 and 20 percent. The operational savings can also justify the resources for an energy program.

Table 4-1 lists examples of some of the operational measures commonly found in manufacturing facilities.

Table 4-1. Operational Measures Example

Measures	Method of identify	Type
Compressed air leakage	Ultra sonic leak dection	Faults
Equipment idling (failed contactor)	Current logging	Faults
Reduce excess air in Boilers	Flue gas Analyzers	Optimization
Oversized motors	Motor load test	Optimization
Variable Speed Drives runnig at 100% speed	Visual inspection of the drive	Faults
Failed Steam trap	Steam Trap Analyzers	Faults
Broken Insulation	Visual inspection	Faults
Clogged Filter	Visual inspection	Faults
Scaling in the condenser	Measurments	Faults
Lights left on all the time (failed controller)	Visual inspection	Faults

Operational measures generally fall into two categories: Optimization of the system and equipment and maintenance issues like faults. There is an increase in energy consumption due to the above problems.

The key challenges in the operational savings is implementing the measures and ensuring that such savings are sustained.

Figure 4-1. Lighting Controller Bypassed

Some recommendations for sustained energy savings include the following:

1. Getting support from senior management;
2. Forming an energy team;
3. Identifying list of operational measures;
4. Installing sub-meters to monitor and track equipment energy consumption on a regular basis;
5. Training operators to operate equipment in an efficient and productive way;
6. Regular testing of steam traps; and
7. Rewarding employees for good behavior.

ENERGY TRAFFIC LIGHT PROGRAM

In the energy traffic light program, the energy-consuming equipment is separated into three categories. There is critical

Energy Analysis

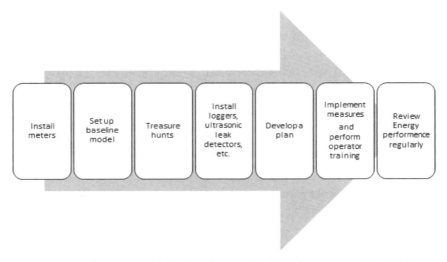

Figure 4-2. Implement and Monitor Energy Savings from Operational Measures

equipment that needs to run all the time, and these are identified and tagged with red markers. Some equipment can be turned off under certain conditions, and these are identified with yellow markers. A third category of equipment can be turned off when not in use, and these are identified with green markers.

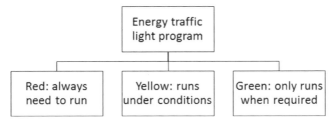

Figure 4-3. Energy Traffic Light Program

COMMON OPERATIONAL MEASURES

Case studies of some common operational measures are given in the following sections, and they show how these operation energy savings are quantified.

Install Occupancy Sensors in Cafeteria

In a food and beverages facility, most of the lights in the cafeteria and in some storage areas are left turned on even when the area is unoccupied. An occupancy sensor can switch off lights when the cafeteria is unoccupied and switch them back on when there is someone in the room. An occupancy logger was installed to monitor unoccupied hours. It was calculated that installation of an occupancy sensor would reduce electricity consumption each year by 11,040 kWh.

Figure 4-4. Occupancy Logger

Reduce Equipment Idling

The plant has different sizes and types of packaging machines for packaging different types of coffee. Some of the packaging machines operate for three shifts while others operate for two shifts. Twelve of the packaging machines have vacuum pumps for vacuum packing. The nominal capacity of various vacuum systems ranges from 490 CFM to 180 CFM, maintaining a vacuum of 29 inches of mercury.

Energy Analysis 143

The following energy conservation options were identified in packaging machines and vacuum pumps:

Reduce Idling of Packaging Machines

The amp for two packaging machines (GL 12 and GL 24) and the associated vacuum pumps were logged for three days.

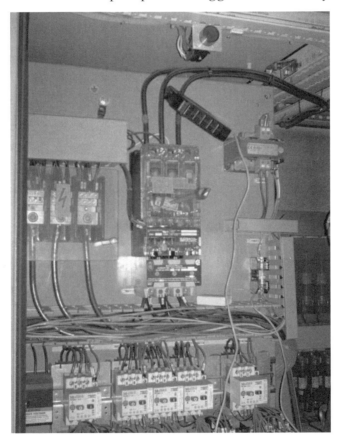

Figure 4-5. Logging of Packaging Machines

The logged amps for one of the GL 12s is shown in Figure 4-6.

It was observed that the vacuum pump was offline while the machines were still drawing power. This indicates that the packaging machines were not actually in production while drawing

144   Industrial Energy Management Strategies

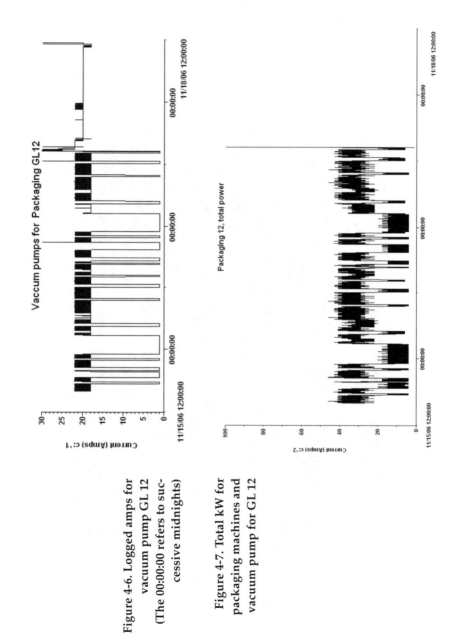

Figure 4-6. Logged amps for vacuum pump GL 12 (The 00:00:00 refers to successive midnights)

Figure 4-7. Total kW for packaging machines and vacuum pump for GL 12

amps. Similar observations were made for GL 24. Some of the heaters in the packaging machines would still be required to be online to maintain temperature for quick startup.

This measure would identify and separate critical loads, like heaters and non-critical loads, like conveyors, and shut off the non-critical loads for each machine when the machine is not in production.

For 5 kW of idle power, out of which 3 kW is the necessary heater load, 2 kW per machine can be saved when it idles. This reduction would minimally affect the peak load of the plant because the machines would not be idling simultaneously. For 12 machines that idle 6.5 hours per day, the savings would be 2 kW X 12 machines X 6.5 hours/day X 300 days = 46,800 kWh.

It is proposed that each machine be equipped with an "idle" switch for the operator to activate. The idle switch will turn "off" all non–essential loads while permitting quick start-up.

Reduce Idling of Vacuum Pumps

The vacuum pumps are operated manually whenever a packaging machine is online. When vacuum pumps idle, they pull air through a spring loaded valve to keep from overheating. This results in very little energy reduction. Analysis of the load profile of one of the vacuum pumps shows that it idles for about six hours per day, and 15 kW (20 amps) was measured during idling.

Providing machine interlocks of the vacuum pump with packaging can reduce the idling time and associated energy waste. In addition, maintenance can be reduced in proportion to reduced run-time. For the measured vacuum pump, the energy consumption could be reduced by 15 kW X 6 hours of idling time/day X 300 days = 27,000 kWh/year.

For all 12 vacuum pumps, if we assume a similar amount of idling time, we can save a total of: 27,000 kWh X 12 = 324,000 kWh.

Intake of Cold Air for Air Compression

The inlet air temperature in relation to the compressor is important because a rise in temperature would increase air volume,

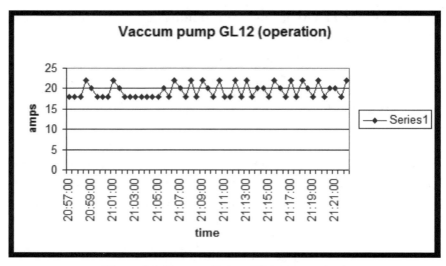

Figure 4-8. Vacuum Pump GL 12 (operation)

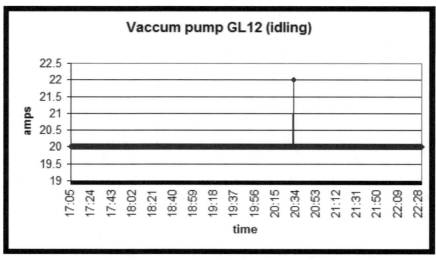

Figure 4-9. Vacuum Pump GL 12 (idling)

which would increase energy consumption. Thus, the lower the air temperature at the compressor inlet, the lower the energy consumption of the compressor.

An energy study of a compressed air system found that the

Energy Analysis

compressors' small air intake pipes and filters are located in the compressor room. The temperature of the inlet air compared with the compressor was higher than the ambient temperature. The actual temperature of air suction (inside the room) for the four compressors is in Table 4-2.

Table 4-2. Air Temperature for Compression Suction (inside the room)

Compressor No.	6	7	2	4
Ambient temperature (°F)	50.2	50.2	50.2	50.2
Suction air temperature (°F)	88.1	88.4	91.6	91.9
Difference (°F)	37.9	38.2	41.4	41.7
% Savings	9.5%	9.5%	9.5%	9.5%

The temperature of air suction in the compressors is well above the ambient temperature. However, care should be taken against an increase in the suction air temperatures.

Lowering the air temperature at the compressor inlet would lower the annual energy consumption of the compressor by 88,862 kWh. This is based on the rule of thumb that for every 10°F reduction in temperature, the energy consumption in the compressors reduces by about 1.9%.

The following measures were identified for the improvement of performance in the compressors:

- Suction air is to be maintained at a temperature close to the ambient temperature for compressor operation in all seasons.

- Relocate the air inlet (suction pipe with filter) of compressors #6, #7, #2, and #4 to outside the room to allow the suction air (temperature and flow) to be ambient.
  — Reduction in electricity consumption is achievable in each compressor. The temperature of the air supplied to the suction of the compressor should be monitored periodically to ensure it is at least equal to the ambient temperature.

Segregate Low- and High-pressure Users of Compressed Air

A fabrication plant requires high-pressure (about 100 psi) compressed air during intermittent periods, though the plant can operate with a lower pressure point of around 90 psi when the fabrication plant is not in operation. Therefore, segregation of high- and low-pressure users can help optimize compressed air operation. Moreover, it would improve system reliability. The analysis indicates that the plant would save 109,924 kWh annually. The reduction in energy due to lowering of pressure can be calculated by the following equation.

FR = Calculated ratio of proposed power consumption to current power consumption based on operating pressure, no. of units.

FR = [((Pdp/Pi) ^[(N X (k-1)/k] -1) − 1]/[((Pdc/Pi) ^[(N X (k-1)/k] -1) − 1]

Where
FR = K/L
FR = calculated ratio of proposed power consumption to current power consumption based on operating pressure, no. units.
K = [((Pdp/Pi) ^[(N X (k-1)/k] -1) − 1]
L = [((Pdc/Pi) ^[(N X (k-1)/k] -1) − 1]
Pdp = proposed discharge pressure
Pdc = current discharge pressure
Pi = 14.7
N = number of compression stages
K = ratio of specific heat =1.4 for air

Optimize sizing of motors

Electrical motors consume about 70 percent of energy in a manufacturing facility. They provide power to equipment like pumps, fans, compressors, etc. Often, motors of higher rating than required for an application are used, which results in poor operating efficiency. Replacing these motors with lower capacity

Energy Analysis

Table 4-3. Compressed air pressure reduction

Proposed Discharge Pressure (gauge pressure)	85.0
Proposed Discharge Pressure (absolute pressure)	99.7
Current Discharge Pressure (gauge pressure)	100.0
Current Discharge Pressure (absolute pressure)	114.7
K	0.7
L	0.8
K/L	0.9
Air capacity available after seperation of air compressors for high pressure users	720.0
kW/CFM	0.2
Reduction of kW due lowering pressure	15.3
Run hours	7,200.0
Electricity Savings (kWh)	109,925.0

motors can improve loading and improve operating efficiency. The following case presents an opportunity to optimize motor size and reduce energy.

Steam Pressure Optimization

A plant has three steam boilers; steam is generated at 100 psi in the boiler, and the pressure is reduced at three stations from 100 psi to 30 psi. Most of the steam is used for space heating. It is used primarily in the make-up air units and the unit's heaters. Some of the steam is used in the waste water treatment process.

The boiler operates at 100 psig. The boiler is oversized, as some of the initial steam load was reduced. The boiler pressure can be lowered to save on gas consumption. The pressure to which it can be lowered will depend on a number of things, including the actual rated pressure. The boiler pressure can be automatically lowered in the shoulder season by automating the pressure switches based on outside air temperature.

The pressure downstream of the PRV can be reduced from 30 psig to 15 psig to take advantage of the increase in latent heat. The plant has two pilot-operated PRVs. For inlet 90 psig and

Table 4-4. Optimize sizing of motors

A.	Motor details	
Rating	74.6	KW
Full load current	93.7	A
Full load speed	1,780	rpm
Full load efficiency	93%	

B.	Existing operation	
Input power	21.4	
Amp	35.1	
Efficiency	87%	
Percentage loading	25%	
Load at shaft *Rated power * loading(%)*	18.7	

C.	Proposed motor	
Rating	30	kW
Full load current	38	A
Full load speed	1,780	RPM
Power loading (for the same load)	62%	
Efficiency (at 50 % load)	93%	

D.	Energy savings	
Reduction in input power *Shaft load * ((1/Efficiency $_{Existing}$) - (1/Efficiency $_{New}$))*	1.31	kW
Energy Reduction (KWh)	10475	KW

outlet 30 psig, the steam flow rate max will be 5,000 lbs./hour each. By reducing the pressure, the capacity of the valve would not be lost. The latent heat at 30 psig is 929 Btu/lb. of steam. The latent heat at 15 psig is 945 Btu/lb. of steam. This is an increase of 1.7%, which can result in an equivalent decrease in steam usage.

Energy Analysis

Figure 4-10. Optimization of Steam Pressure

VERIFICATION OF OPERATIONAL SAVINGS

The steel plant carried out plant metering to identify some operational measures. The operational measures were implemented and an action log for various measures was developed. A CUSUM analysis was carried out on the electricity data obtained from the main interval meter, and it showed a reduction of about 436,000 kWh. RETSCREEN Expert was used to carry out the CUSUM analysis. In addition, the action logs of the measures were carried out. The action logs are shown in Table 4-4. The savings for different measures have been calculated from the CUSUM analysis.

Table 4-5. Calculation of Operational Savings

PROJECT TYPE	PROJECT DESCRIPTION	DATE IMPLEMENTED	PROJECT TYPE	COMMENTS
OPERATIONAL SAVINGS	Cooling tower fan motor rectified	30/12/2011	Operational	The cooling tower had three cell and all three fan were running irrespective of the demand. Rectification of control help in shutting down one motor
OPERATIONAL SAVINGS	Compressed air leak	30/4/2012	Operational	Compressor leaks were repaired
Optimization	Cooling water system pump shut down	30/9/2012	Optimization	Piping modificaion was done and one of the pump was shut off

Energy Analysis 153

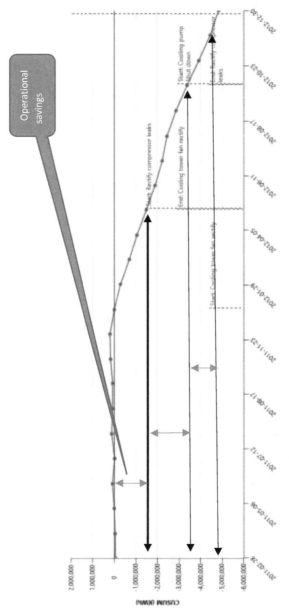

Figure 4-11. CUSUM Analysis for the Steel Plant Energy Optimization

154   Industrial Energy Management Strategies

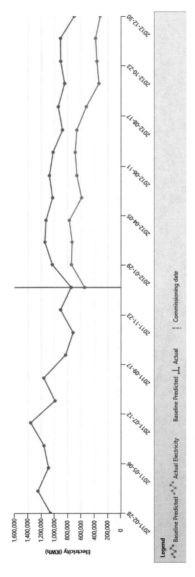

Figure 4-12. M&V for Steel Plant Energy Reduction

Chapter 5

Selling Energy Projects

Despite having significant energy efficiency potential, several key barriers hinder the ability to achieve energy savings at industrial facilities. These barriers include:

- End-users often lack the capital budget to self-finance energy efficiency investments.

- Efficiency projects at industrial facilities have to navigate an investment decision-making framework that places a heavy emphasis on optimizing manufacturing processes and ensuring continuous operation of plant assets.

- A high ratio of transaction costs (i.e., conducting preliminary and detailed audits and establishing M&V protocols) to total project costs hinders the cost effectiveness.

- Corporate capital budgeting processes place energy efficiency in direct competition with other core priorities, such as investments that expand production, increase throughput, and maintain overall plant reliability.

- Industrial firms have a short-term horizon for investments and typically require that projects have rapid payback periods.

- Lack of internal human resources to identify, execute, and verify energy conservation or efficiency projects.

The energy champion would need to address some of the above barriers and sell their projects to senior management. Following are some tips that might help in selling the project to management and getting the necessary funding. One of the key issues to getting more projects approved is understanding energy and all related issues of the facility and connecting all the relevant issues the project might be able to address. In summary, for successful selling, it is important to enlist and quantify all types of benefits from the project and optimize on the project costs.

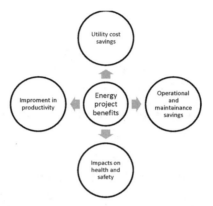

Figure 5-1. Benefits from an Energy Project

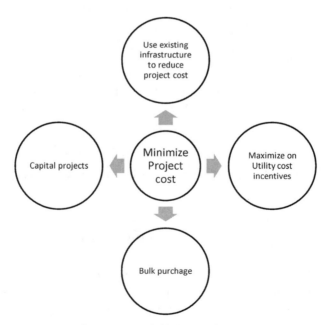

Figure 5-2. Minimize Project Cost

USING THE CORRECT MATRIX

Several energy indicators can be used, such as payback period, net present value, and internal rate of return. It is important to understand the advantages and disadvantages them, and below are the definitions of some project financial matrixes.

> SIMPLE PAYBACK = Project Investment/ Monetary value of cost savings

Net present value (NPV) sums up the present value of all future cash flows (net). Therefore, this matrix quantifies the net benefit that would occur throughout the project's life.

> PV (PRESENT VALUE) = CASH FLOW /(1+ discount rate)

Internal rate of return (IRR) for a project is the value of the discount rate at which the net present value is zero. The savings-to-investment ratio is the ratio between the present value of the cash inflows and the present value of the cash outflows for the complete project life.

> Savings to investment ratio = PV of all cash inflows/present value of all cash outflows

In one of the plastic processing facilities, a 300-ton chiller broke down one compressor stage. The energy manager of the plant conducted an analysis for two options: repair the existing chiller or buy a new chiller. Tables 5-1 and 5-2 show yearly energy and maintenance savings discounted using a discount rate of 10 percent. Note the investment on the new chiller, which is a cash inflow negative, compared to project savings. The various project matrixes are calculated for the chiller replacement option; similar analysis is also carried out for the repair option. Even though the repair option has a much lower payback period, the replace option has a much higher net present dollar value comparably.

Use All Free Money Available for the Project

Because the financial matrix of an energy reduction project depends on the project's costs and savings, it is important to maximize all types of incentives to reduce costs. The project's financial performance improves greatly if an analysis captures both utility and non-utility financial benefits.

Replacement of aging equipment is an excellent opportunity to incorporate energy efficiency into the system. In a plastic extrusion plant, two projects were prioritized for implementation: The compressed air retrofit and the blower optimization. The compressed air retrofit project involved the installation of a VFD air compressor. The savings for the project was calculated to be 90 kW demand and 430,000 kWh. Because the compressor was almost at the end of its useful life, the energy manager focused on the compressed air system. The company had already approved the cost of a new compressor, so he engaged a consultant to review the complete system. It was found that the compressed air system had leakage. A plan was in place to identify and repair the air leaks, which allowed additional savings and also helped in optimizing the size of the new compressor. The energy manager also applied for a utility incentive. Lower compressor cost and savings from the leak repair enabled funding for the blower optimization project.

Take a systems approach and bundle all opportunities together to improve overall financial performance of a project.

Selling Energy Projects 159

Table 5-1. Repair Failed Chiller

Tons	300
Existing kW/ton	1.2
New kW/ton	1
Savings	60
Energy Savings	420000

Cost of Chiller ($)	$30,000
Discount Rate	10%
Energy Rate ($/Kwh)	0.1
Annual Energy Savings (KWh)	420,000
Energy Savings ($)	$42,000
Maintaince Savings	3%

Year	0	1	2	3	4	5	6	7	8	9
Chiller Cost ($)	-$30,000						-$300,000			
Energy Savings ($)		$42,000	$42,000	$42,000	$42,000	$42,000		$126,000	$126,000	$126,000
Maintence Savings($)		$12,600	$12,600	$12,600	$12,600	$12,600		$3,780	$3,780	$3,780
Total	-$30,000	$54,600	$54,600	$54,600	$54,600	$54,600	-$300,000	$129,780	$129,780	$129,780

Present Value of Cash inflows ($)	-$30,000	$49,636	$45,124	$41,022	$37,293	$33,902	-$169,342	$66,598	$60,543	$55,039
Present value of cash out flows ($)							$0			

NPV ($)	$189,815
Simple Payback Period (yrs)	0.7
Savings to Investment ratio	2.0
Return on Investment	25%

Table 5-2. Replace Failed Chiller

Tons	300
Existing kW/ton	1.2
New kW/ton	0.6
Savings (kW)	180
Energy Savings	1260000

Cost of Chiller ($)	$ 300,000
Discount Rate	10%
Energy Rate ($/Kwh)	0.1
Annual Energy Savings (KWh)	1,260,000
Energy Savings ($)	$ 126,000.00
Maintaince Savings	3%

Year	0	1	2	3	4	5	6	7	8	9
Chiller Cost ($)	-$300,000									
Energy Savings ($)		$126,000	$126,000	$126,000	$126,000	$126,000	$126,000	$126,000	$126,000	$126,000
Maintaince Savings($)		$3,780	$3,780	$3,780	$3,780	$3,780	$3,780	$3,780	$3,780	$3,780
Total	-$300,000	$129,780	$129,780	$129,780	$129,780	$129,780	$129,780	$129,780	$129,780	$129,780

Present Value of Cash Inflows ($)	-$300,000									
Present value of cash out flows ($)		$117,982	$107,256	$97,506	$88,641	$80,583	$73,257	$66,598	$60,543	$55,039

NPV	$447,406
Payback Period (Yrs)	2.4
Savings to Investment ratio	2.5
Return on Investment	43%

Selling Energy Projects 161

Figure 5-3. Replace Failed Chiller (RETSCREEN Expert)

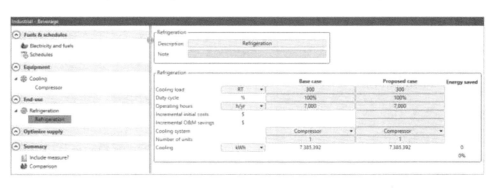

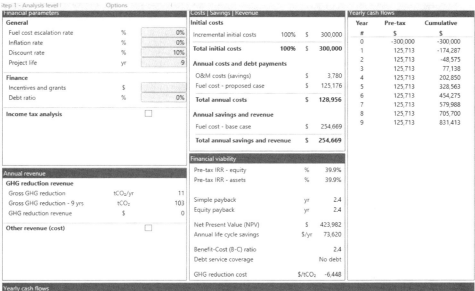

Figure 5-4. Replace Failed Chiller (RETSCREEN Expert)

The following case study illustrates that bundling energy savings improves payback for projects.

In another manufacturing facility, the cooling water system had an inefficient pump. The energy manager of the facility conducted a detailed system study that identified additional energy savings by shutting down another pump with minor modifications in the piping system. The two cases were compared and the one that included an efficient pump with piping modification to shut off the pumps showed better financial performance.

CASE STUDIES ILLUSTRATE OPPORTUNITIES FOR MULTIPLE UTILITY SAVINGS

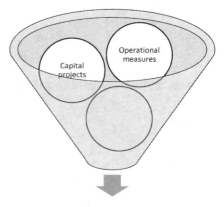

Decrease in payback period

Figure 5-5.
Bundling capital and operational measures

Some processes use multiple utility streams, such as gas, electricity, and water. These systems often present opportunities to reduce operating costs for more than one utility. The aggregation of all the cost and energy reduction opportunities often helps in improving the financial performance of the project. The following case discusses such an opportunity in detail.

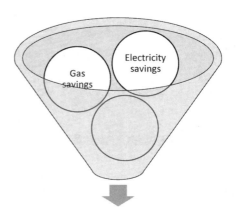

Reduces payback

Figure 5-6
Savings from Multiple Utility Streams

Selling Energy Projects 163

Table 5-3. System Approach Improves Financial Performance

Base Case	
Inefficient cooling pump (KW)	52
Total Load (KW)	52
Total Energy (kWh)	403,104

Proposed Case	
Efficient cooling pump (KW)	22
Total Load (kW)	22
Total Energy (kWh)	170,544

Savings	
KW	30
kWh	232,560
Project cost	$50,000
Savings	$30,233
Payback	1.65

Base Case	
Inefficient cooling pump (KW)	52
Process pumps load	137
Total Load (KW)	189
Total Energy (KWh)	1,466,678

Proposed Case	
Efficient cooling pump (kW)	22
Process pump load after modification (kW)	90
Total Load (kW)	112
Total Energy (kWh)	868,224

Savings	
KW	77
kWh	598,454
Project cost	$60,000
Savings	$77,799
Payback	0.8

Water Pumping System

City water is stored in an underground tank and pumped to different areas. The water is collected, treated, and discharged into the lake. A study of the system reveals several opportunities to reduce operating costs and energy consumption in the water system.

Complete Transition to Closed-loop Cooling

The production process was using close to 421,228 cubic meters per year of "once-through" water for cooling the extruder bath. The water from the extruder bath was treated before it was discharged into the lake. Because the existing cooling process was costly, it was proposed to use recirculated water through a cooling tower. The transition from once-through water to recirculated water would also reduce the city water requirements.

Estimated Savings

Data from the monthly utilities report were used to estimate annual water consumption and costs. In the closed-loop cooling system, some water would still be lost to evaporation from the cooling tower. To estimate how much water is supplied to the cooling tower, it was assumed that the average temperature rise of water in the extruder bath is about 20°F, and that the average temperature drop of water through the new cooling tower would be about 15 degrees. In the region, about 5 percent of cooling tower water is generally lost to evaporation. Annual pumping costs for the cooling tower can be estimated from the rate of water delivered to the cooling tower. A motor/pump could deliver the required flow rate. One pump would be needed for the supply to the cooling tower, and one pump would be needed for the return. The estimated cost of implementation was about $100,000.

Lower Running Time in the Water Treatment Process

The water that passes through the extruders is treated before it is discharged into the lake. There would be a reduction in the volume of water to be treated once a recirculation system is used in extruder bath cooling. Therefore, due to a reduction in the volume of water to be treated, the waste water treatment pump's operating time would be reduced drastically.

Energy Saving Calculation

Based on the assumption that about 70% of the water is used in an extruder bath, then a similar reduction can be assumed in the water treatment and, accordingly, about the same reduction in the electricity consumption in the water pumping system.

Implement Low Cost Project

Low cost opportunities can help demonstrate the value of operational energy savings to senior management. In addition, low cost opportunities can be funded from a maintenance budget. Installing meters on equipment can help in quantifying losses. The business case developed using information from energy meters is more credible and realistic. In addition, the data can be used to develop a baseline for carrying out measurement and verification (M&V) for the implemented project. Data regarding other benefits from the project, such as reduction in maintenance hours, need to be captured.

Similarly, once the project is implemented, the post-project information regarding energy consumption and other benefits needs to be captured. This information would help show the true value of the project to all stakeholders. This type of analysis would help get funding for the next energy project.

Table 5-4. Water Saving calculations

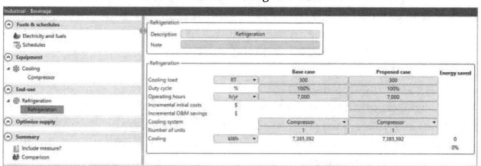

Selling Energy Projects

Table 5-5. Energy Saving Calculations for Water Treatment Pump

Existing horsepower of waste water treatment pump	30	
Load factor	60	%
Annual hours of operation	8,000	hours
Reduction in run hours of the pump	0.7	
Electricity savings	75197	kWh
Cost savings	$5,715	
Investment	$8,000	
Payback	1.4	

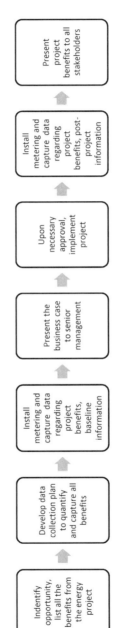

Figure 5-7. Low Cost project Implementation Stages

Chapter 6

Measurement and Verification of Industrial Energy Conservation Projects

INTRODUCTION

The measurement and verification (M&V) process can help in gaining support from all stakeholders towards implementing a sustainable energy management program in an industrial facility. There are other benefits, like increasing energy savings, enhancing financing for energy efficiency projects, and educating all facility users about their energy impacts. M&V is also an important component for the ISO 50001 Standard. From the perspective of the incentive program, the M&V process ensures that the incentive dollars are used effectively to achieve the desired objective of reducing energy.

CASE STUDIES

Resources (primarily skills) required to carry out the M&V process are limited. Some energy efficiency/conservation measures implemented in industries can be very complex, and may require activities like process modeling and simulation to carry out M&V. Organizations like AEE and EVO are playing key roles in addressing the skill requirements in this area. Case examples of measurement and verification (M&V) carried out in small- and mid-sized manufacturing facilities are presented.

Table 6-1. IPMVP Options

Options	Description	Examples
A Partially Measured Retrofit Isolation	Savings are determined by partial field measurements of the energy use of the system to which an ECM is applied. Some, but not all parameters may be stpulated	Lighting retrofit where pre - post retrofit fixture wattage are measured, operating hours of lights are generally agreed upon
B Retrofit Isolation	Savings are determined by field measurements of the energy use of the system to which the ECM was applied	Variable speed drive on a pump, electricity use is measured by a kWh meter installed on the electrical supply to the pump motor
C Whole Facility	Savings are determined by measuring energy use at the utility meter level. Bills may	Several ECM affecting many systems in a manufacturing facility, Utility bill (main utility meter) are used.
	be corrected for production, weather etc	
D Calibrated Simulation	Savings are determined by using building simulation. This options is rarely used and is used primarily when there is no pre -retrofit Utility data available	

Case Study 1

This case is taken from a plastics molding plant. The hot water from the molding machines is stored in the main tank. The hot water is then passed through a heat exchanger where it is cooled, and the cold water is then stored in the process tank. The water from the process tank is pumped back to the molding machines by the process pumps. In the base case, three out of the four pumps operate. The hot water in the secondary side of the heat exchanger is cooled at the cooling tower. Similarly, three process pumps are in operation in the base case. Flow measurement on all three pumps revealed that the existing pump has enough capacity so that the flow requirements can be met with just two

Measurement and Verification of Industrial Energy Conservation Projects   171

of the three pumps. It was also found that the heat exchanger pumps are oversized and not operating in the BEP best efficiency point. Additionally, they are controlled with the help of manual valves. This is not an efficient form of control. The retrofit project involved installation of a new heat exchanger pump with VFD and shutoff of one process pump.

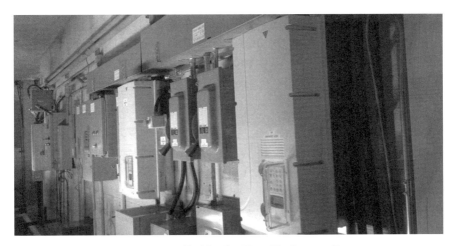

Figure 6-1. Installed in the Heat Exchanger Pump

Figure 6-2. Flow Measurements for Pump Optimization

As per the measurement and verification (M&V) plan, it was proposed to follow International Performance Monitoring and Verification Protocol (IPMVP) Option B, which includes all parameter measurements. It was decided to install loggers to measure pump kW and kWh for all the pumps. The total system flow was also measured (for normalization). The efficient case measurements also involved measuring the total kW and kWh for each of the pumps. The total system flow was measured, it was found to be same with the base case, so no adjustment was carried out. Therefore, the kW and kWh savings were calculated as 90 kW and 136,655 kWh.

Case Study 2

A commercial laundry was retrofitting the dryer fan with a more efficient system. There were 20 fans. In the base case it was decided to take spot measurements of amperage on each of the 20 fans. Since the load on the fans was fairly constant, a sample size of 20% was taken for detailed study. Amp loggers were installed for a week.

The details of the dryer loading were noted for normalizing the same in the efficient case. In the efficient case, spot measurements were taken on all 20 fans, and detailed study was carried

Figure 6-3. Loggers Installed for Measuring the Amps of the Dryer.

out on 20% of the fans. Again amp loggers were installed for a week. Details of the dryer loading were noted. It was observed that there was no noticeable difference in the dryer loading. Therefore, the kW and kWh savings was calculated as 123 kW and 984,000 kWh.

Case Study 3

A stationary product manufacturing plant decided to implement low-cost and operational measures. Some of the low-cost measures included the installation of motion sensors and business automation system (BAS) recommissioning. In addition, an energy awareness campaign was launched at the company's health and safety meeting. Very specific measures, like shutting down conveyors during lunch hours, were discussed. It was important to measure progress and demonstrate results of the initiatives to senior management.

As per the measurement and verification plan, it was proposed to follow International Performance Monitoring and Verification Protocol (IPMVP) Option C, since the measure involved the whole facility. Monthly production data were collected with the corresponding energy consumption for the same period. The energy consumption was based on the main utility meter.

A linear regression between monthly production and energy was developed and we found a high regression coefficient (R^2=0.78). This was used as a baseline model (y=0.0206X+7658), which is in the form Y=MX+C. The regression model was then used as a basis for evaluating monthly energy performance for the facility. CUSUM analysis was carried out and an electricity savings of 218,048 kWh was identified.

Case Study 4

The fourth case is taken from a lighting retrofit project at a brewery. The project involved replacement of 400-watt metal halide lights with T5 banks and T12 fluorescent lamps with T8 lamps and electronic ballasts. As per the measurement and verification (M&V) plan, it was proposed to follow International Performance

174 Industrial Energy Management Strategies

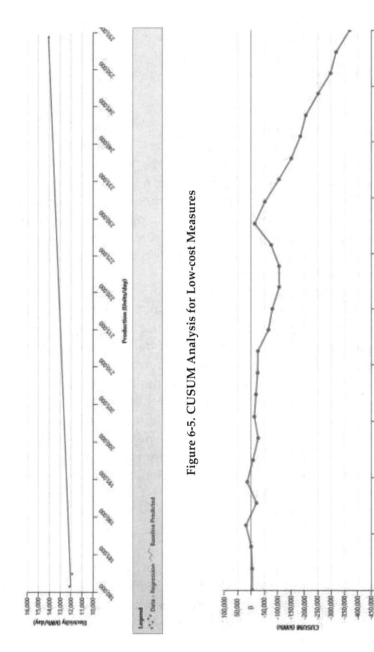

Figure 6-4. Regression Analysis

Figure 6-5. CUSUM Analysis for Low-cost Measures

Measurement and Verification of Industrial Energy Conservation Projects 175

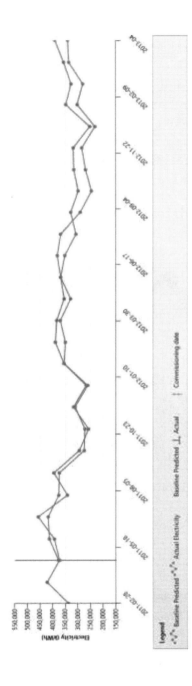

Figure 6-6. M&V for Low-cost Measures

Monitoring and Verification Protocol (IPMVP) Option A, which includes key parameter measurement. In the baseline period, kW was measured for the existing 400W metal halide and fluorescent lamps. The reporting period measurements consisted of measuring the kW of the new fixtures installed. The lamp burn hour was to be provided by plant personnel. The brewery did not have an energy management program, so measurement and verification of the savings was not considered as an essential component in the project implementation. We faced several hurdles during the verification of the energy savings, such as identification of electric circuits for retrofit fixtures.

The total kWh from the lighting retrofit project was verified to be 400,201 kWh. The demand reduction was verified to be 70 kW.

Figure 6-7. Lighting Retrofit Project

Case Study 5

The case study consists of installing lighting controls, namely photocell and occupancy sensors, in some of the lighting fixtures in the brewery. As a part of the M&V plan, kW and kWh were to be measured for the existing and retrofit fluorescent lamps for a day, both for the baseline and reporting period. As mentioned, the plant did not have an energy management program, and the tasks of identifying and locating the circuits for installing the data

Measurement and Verification of Industrial Energy Conservation Projects 177

loggers primarily in the reporting period were a challenge, as the project manager did not consider the M&V process an important part of project implementation. The total kWh from the lighting control project was verified to be 38,073 kWh.

Figure 6-8. Motion Sensors Installed in a Brewery

Case Study 6

The case involves installation of a cooling water bypass pump in a cogeneration plant. The cooling water pump is a part of the cogeneration plant and is driven by a 500 kW motor. The cogeneration plant runs for about 18 hours a day, while the cooling water pump runs for 24 hours a day. As part of the energy conservation measure, the plant decided to shut down the large pump whenever the cogeneration plant was not running. It was decided that a smaller pump with a variable frequency drive would be installed to supply a minimum flow in the cooling water system to inhibit corrosion and biological growth.

Two projects were implemented. The first project was for demand reduction, while the second was for energy reduction (kWh).

The daily variation of amps from the distributed control system (DCS) was noted, to study the load pattern for the existing pump. Other parameters, like the position of the valve, were also noted.

Figure 6-9. Cooling Tower for the Cogen Plant

It was also mentioned by the operator that valve position was not generally changed. Based on the information collected from the site, it was proposed to follow IPMVP Option A, which includes key parameter measurement. In the baseline period, kW was measured for the existing pump. The run hours were noted from the DCS system.

The reporting period measurements also consisted of measuring the kW of the newly installed pump, and the run hours were noted from the DCS system.

The cogeneration plant had an energy management program, so there was good cooperation during the M&V review process of the pre- and post-project implementation phases. Additionally, they had a data acquisition system that provided critical information for the measurement and verification (M&V) process. One of the other challenges was measurement of power of the high-voltage motor that was driving the pump. It was decided to take the motor reading from the DCS system. The total demand and electricity savings from the projects was verified to be 335 kW and 2,297,683 kWh.

Case Study 7

The case is taken from a coffee packaging facility. There are three air compressors in the plant. Compressor 3 is a 75 HP lubri-

Measurement and Verification of Industrial Energy Conservation Projects 179

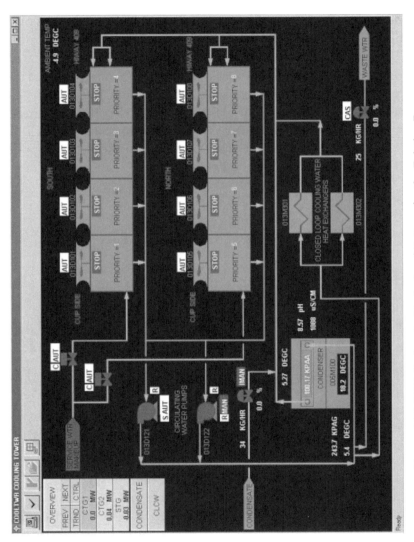

Figure 6-10. DCS System to Study Load Pattern for the Existing Pump

cate screw compressor. Compressor 4 is a 50 HP vane compressor, and compressor 5 is a 30 HP vane compressor. Each screw compressor has its own refrigerated air dryer. The compressed air is collected in a receiver before it goes to the plant. The compressed air is supplied by the compressor banks and used for pneumatic controls of the packaging machines. Amperage data for all the operating air compressors, namely compressor 4, compressor 5, and compressor 3, were logged to determine the load profile. The energy retrofit project involved installation of compressed air sequencers. As per M&V plan, it was proposed to log the amperage of all the compressors in the base case and determine the ACFM using Air Master (US DOE ENERGY TOOL). In the reporting period, it was proposed to log the amps.

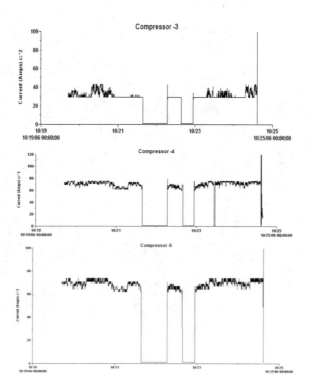

Figure 6-11. Amp Measurements Carried out at the Compressors

A regression model was created to correlate the number of machines that were being operated by the air compressors. In the reporting period, the compressors were operated with the sequencer, and the amperage of the compressors was logged to determine the kWh. The kWh of the baseline period was adjusted to the production level of the reporting period using the regression model to calculate the savings. The electricity and demand savings were verified to be 31,260 kWh and 6 kW.

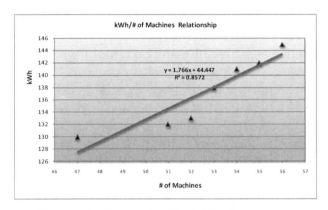

Figure 6-12. Regression Model for Compressors

The coffee packaging facility had an industrial energy management program, and cooperation from the plant enabled an exhaustive M&V process. The process also involved interdepartmental cooperation within the facility, and gave production data for the same. The M&V process was also very educational for facility personnel, and all involved shared the success of implementing the compressed air sequencing project.

Case Study 8

Floating head pressure was implemented on three 30-HP refrigeration compressors that serve the freezers. An automation system controls the operation of the refrigeration compressors and the system was operated at 250 psig condensing pressure. The floating head controls the condensing pressure based on outside air conditions. Variable speed drive installed in the condenser fan helps in varying the condensing pressure.

As per the M&V plan, it was proposed to monitor the electricity consumption of the condenser and the compressors in the baseline; in the efficient case, it was proposed to conduct an ON-OFF test wherein the compressor pressure varies between on and off conditions for a week. This test would occur during summer, winter, and spring. The savings would be averaged and applied to the annual base case consumption.

Power monitors were installed in the compressors and condensers, and pressure and temperature data were obtained from the automated system. Monitoring occurred for six months to obtain the energy consumption in the compressors for the representative sample. After the project was installed and commissioned, the ONOFF test was carried out in the compressors. In the ON state, the compressor was allowed to operate the system with floating head pressure, and in the OFF state, the floating head controls were bypassed. The results of the measurement and verification are given in Table 6-2. The savings from the ON-OFF test for winter months is 30% and average reduction is 15%, which results in a savings of 127,841 kWh.

Case Study 9
Chilled Water Plant Optimization

This project involved optimization of the refrigeration system in a cold storage. Specific measures considered were:

- Installation of VFD on the evaporator fans
- Installation of floating head pressure control on some of the freezers
- Proper sequencing of refrigeration compressors
- Optimize setpoints on some of the freezers

Per the M&V plan, IPMVP, Option C was proposed to verify savings. It was proposed to use utility bills for the analysis. The

Measurement and Verification of Industrial Energy Conservation Projects 183

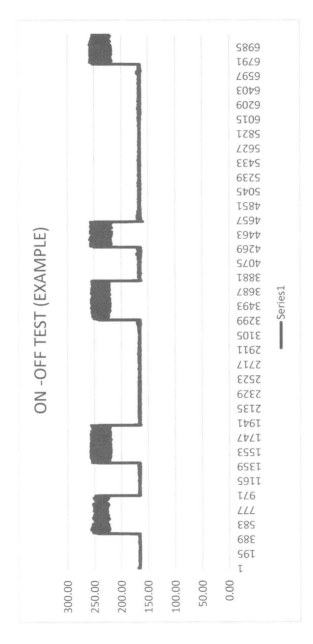

Figure 6-13. ON OFF test: Pressure Profile on the Compressor

Table 6-2. ON-OFF Test Results

Row Labels	Daily KWh consumption	Average of PSI	STATUS
05-Dec	105	166	ON
06-Dec	454	185	OFF
07-Dec	518	222	OFF
08-Dec	458	187	OFF
09-Dec	515	220	OFF
10-Dec	427	166	ON
11-Dec	425	166	ON
12-Dec	462	189	OFF
13-Dec	515	220	OFF
14-Dec	457	186	OFF
15-Dec	477	197	OFF
16-Dec	425	166	ON
17-Dec	425	166	ON
18-Dec	426	166	ON
19-Dec	443	176	OFF
20-Dec	374	237	OFF
Grand Total	6905	189	

estimated savings was greater than 10% of the utility bills. As per the proposed M&V plan, monthly energy consumption was collected from the utility bills and production figures were requested for the same period. Average air temperature and production was used as the independent variable.

Once the new system was installed and commissioned, monthly energy and production data were collected. The difference between predicted baseline and actual consumption showed a savings of 5,141,590 kWh.

Case Study 10
Dust Collection System

The project involved retrofitting a dust collection system for a wood product manufacturing company. The existing dust collection system was used to exhaust wood dust from each wood-

Measurement and Verification of Industrial Energy Conservation Projects 185

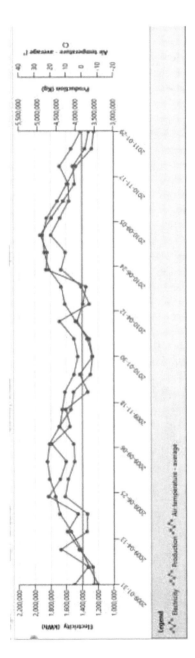

Figure 6-14A. Energy Performance Model

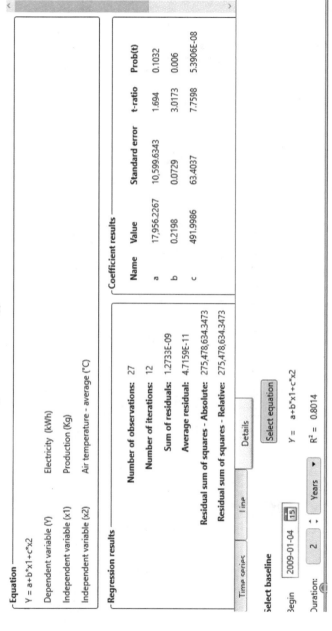

Figure 6-14B. Energy Performance Model

Measurement and Verification of Industrial Energy Conservation Projects 187

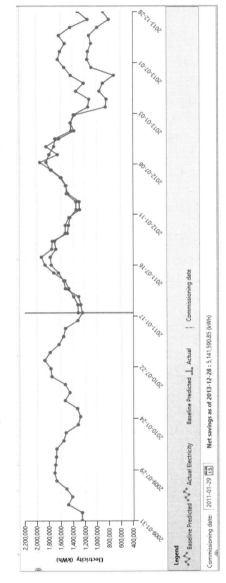

Figure 6-15. Measurement and Verification of Savings

working machine; 35 motorized gates were installed at various work stations. Variable frequency drive was installed in the 100 HP exhaust fan. A programmable logic controller was installed to control the system.

Per the M&V, Option A was used to calculate energy savings. It was proposed to measure daily energy consumption for the exhaust fan for a week. Similarly, during the post-project period, it was proposed to measure daily energy consumption for the exhaust fan with VFD installed for a week.

Per M&V, an amp logger was installed in the exhaust fan for the baseline case and the same info was collected for the post-project period. Details of M&V are given in Table 6-3.

Table 6-3. Energy Savings Summary for Dust Collector

PARAMETERS	BASE CASE	POST PROJECT
VOLTAGE	600	600
AVERAGE AMPS	138	80
POWER FACTOR	0.8	0.8
RUNHOURS	8000	8000
KW	114.72768	66.5088
KWh	917821.44	532070.4
	SAVINGS (KWh)	385751

Reference

Bhattacharjee, Kaushik (2011). Measurement and verification of industrial energy conservation projects presented and published at World Engineering Congress Conference, Chicago.

Measurement and Verification of Industrial Energy Conservation Projects 189

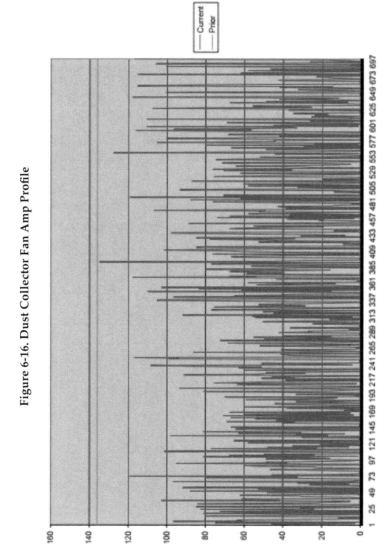

Figure 6-16. Dust Collector Fan Amp Profile

Chapter 7

Project Pitfalls

ENERGY MANAGEMENT AND
ENERGY EFFICIENCY PROJECTS

Companies with a sustainable energy management policy follow a strategic approach toward projects and maximize those projects' value. Some of that value includes the following:

- Energy reduction during complete life cycle of implemented projects;
- Reduction in maintenance costs and utility costs;
- Maximizing utility incentives;
- Reduction in greenhouse gas emissions;
- Minimizing energy waste;
- Demonstration of corporate social responsibility; and
- Optimizing investment on capital upgrades.

As such, these companies follow a systematic approach to developing and implementing projects, whereas other companies might skip some of these stages if they don't have a sustainable energy management policy in place.

PROJECT PITFALLS

A systematic approach following best practices in the project development cycle will allow a company to maximize on the value of its projects.

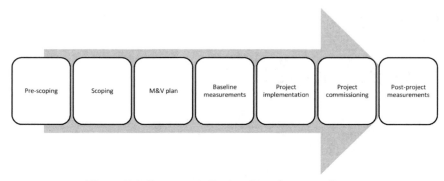

Figure 7-1. Systematic Project Development Stages

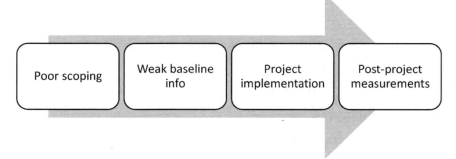

Figure 7-2. Ad Hoc Approach to Implementing Projects

Some of the pitfalls encountered at various stages in a project's development cycle are identified below.

Pre-scoping
Project Prioritization

Few companies have developed a comprehensive list of all issues and opportunities in their manufacturing plants, and as a result, energy project prioritization does not take place. This leads to decreased savings. In addition, the projects implemented might need to be completely overhauled. The below project prioritization spreadsheet lists all opportunities in a facility and ranks them based on specific criteria, such as payback, ease of implementation, capital and operational opportunities, and other criteria. The projects are then prioritized for more detailed study, metering, and other evaluative measures.

Table 7-1. Project Prioritization Spreadsheet

	OPPORTUNITIES	COMMENTS	RANK
1	REPAIR COMPRESSED AIR LEAKAGES	Tag leakage and repair them	1
2	INSTALL VFD COMPRESSORS	Install after repairing leaks	2
3	LIGHTING RETROFIT	Can implmenent lighting retrofit	1
4	INSTALL VFD COOLING WATER PUMPS	Review complete system	2
5	INSTALL NEW CHILLER	Capital project, review complete system	3

Implement Demand Measures before Supply Measures

It is important for a system to implement demand reduction opportunities before looking at such opportunities on the supply side. For example, a food processing facility had five fixed-speed rotatory compressors with a capacity of 4950 CFM. The project involved replacing one existing compressor with variable speed to optimize the system. A detailed system study was conducted and it was found that the system had about 40% air leakage. The leakage was identified and repaired before implementing the project. The project freed up compressed air demand, which resulted in shutting down one of the existing compressors and buying a lower capacity VFD compressor. The project payback improved due to additional savings and lower project cost.

Systems Approach to Generate More Savings

Energy savings can be maximized by taking a systems approach. A plastics manufacturing plant had chilled water via three air cooled reciprocating compressors with a capacity of 100 tons. It was proposed to replace them with a water cooled 300-ton centrifugal chiller with an energy performance of 0.6 kW/ton. Energy analysis of the base case and proposed case showed that it was possible to save 502,531 kWh through the replacement. However, with the addition of a VFD on the chilled water pump and installing a free cooling heat exchanger, the savings would increase to 717,452 kWh.

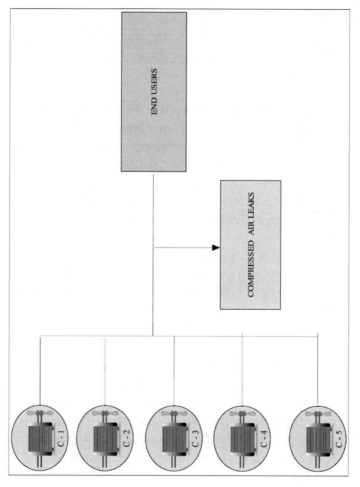

Figure 7-3. Compressed Air with Leakage

Project Pitfalls 195

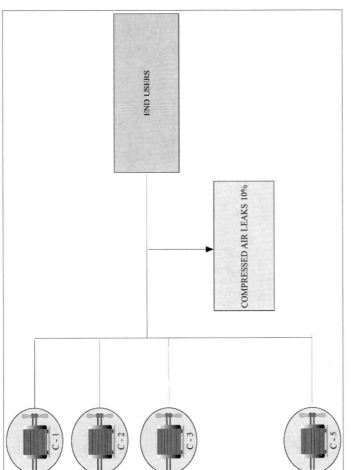

Figure 7-4. Compressed Air System after Repaired Leakages

Pilot Projects

Pilot projects are a good idea for proof of concept. These projects help to demonstrate energy savings while addressing the project's concerns. Often, not enough documentation and research is done before field trials. Field trial results can help during the scoping stages, and many times, trial results can help sell the project to management. Trial design involves a detailed review of the system to determine the baseline and retrofit condition and promote understanding of the objectives of the trial, i.e., whether it is trying to determine savings and/or other concerns, such as impact on the production process, the type of metering required, etc. The cost of metering has decreased substantially, and it is now possible to install low-cost loggers. Sometimes it might be also possible to get information for the DCS or the process control system.

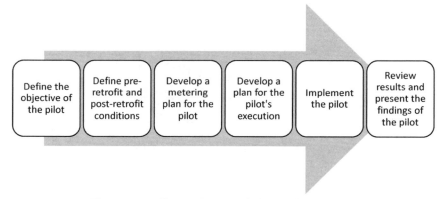

Figure 7-5. Pilot Project Implementation Stages

Scoping

Once projects are prioritized in the pre-scoping stage, it is important to review all possible ways to maximize energy savings, such as the equipment's efficiency rating, the kW/ton of the chiller being considered, insurance that over-sizing of the equipment is avoided, whether the equipment is going to be controlled, and other factors. For example, a chilled water plant can operate at a lower condenser water temperature. One should consider whether it is possible to include some kind of permanent metering to review operation on a regular basis.

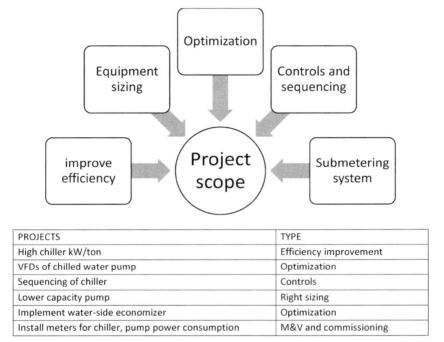

Figure 7-6. Factors for Maximizing Energy Savings

POST-PROJECT OPERATING STRATEGY

The project scoping should also include operational considerations, such as redundancy, and its impact on energy savings. A cooling water system in a metal processing facility was operating one 150-horsepower pump driven by a variable speed motor. The flow of the system was reduced, so the plant decided to install two small 100-horsepower pumps with a VFD, and the second pump would remain as a standby system. For system reliability, the operations team decided to operate both pumps together. The changes in the proposed operation affected the energy savings it reduced. The lesson learned in the project was to include all issues as much possible during the project scoping stage, such as post-project operation, including redundancy consideration.

IMPLEMENTATION SEQUENCE ON AN OPTIMIZATION PROJECT

To maximize energy savings in a centralized system, such as a chilled water or compressed air system, it might be useful to implement the project in phases. The initial phases would focus on installing submeters, upgrading systems (such as controls), and installing VFD on pumps. The later stages can focus more on system optimizations.

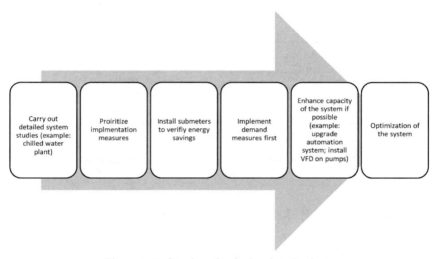

Figure 7-7. Staging Optimization Projects

A large plastics manufacturing company has a cooling water plant that has four chillers with a total of 8,800 kW (2,500 ton) cooling capacity, operating 365 days a year. The chilled water is used for process cooling by three main chilled water pumps. Heat rejection for the chillers is carried out by a six-cell cooling tower.

As part of an initiative to reduce operating costs, the plant installed submeters at some of the large energy consumers, like chillers, pumps, etc. The system was capturing data on an hourly basis. This allowed them to benchmark chilled water performance and also track improvements in terms of lowering energy consumption.

A recommissioning study discovered that more than the required number of chillers were operating and the chilled water pumps were consistently operating at full flow, resulting in less than optimal efficiency of the cooling system. The chilled water flow was controlled by manually operated valves. The annual energy data from the submetering system allowed quantification and verification of savings.

The study identified several necessary changes, including staging the chiller and pump operation, installing VSDs on the chilled water pumps, and replacing manual valves with automatic valves controlled by the BAS.

The chiller optimization project was broken into two phases. The first phase involved installation of VFD on chilled water pumps. The second phase involved replacement of manual valves with an automatic valve, chiller staging and a complete chiller controls optimization.

The CUSUM analysis for the 265 days of the project optimization period is presented in Figure 7-8. As can be seen from the figure, Phase 1 and Phase 2 implementation generated a total savings of about 1.1 million kWh, which is about 23% of the energy consumption for 265 days.

The success of the chilled water optimization project in the plastic plant can be attributed to the systematic approach to implementing the project, such as installation of submeters for all major equipment, carrying out recommissioning studies to identify all opportunities, review of energy improvement on a regular basis by plant management, and operator training. The changes in the chilled water plant operation strategy were first implemented as a pilot project, and the results were monitored; operators were trained during implementation. Detailed documentation on implementation of the chiller operating strategy was prepared by the consulting company for ongoing reference by the building operators.

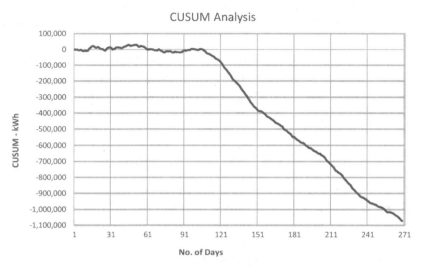

Figure 7-8. Chiller Plant CUSUM Analysis

M&V

Measurement and verification (M&V) might be a project requirement because of utility incentive, but it is also important to show value to all stakeholders. It may happen that the project economics will be better than those presented in the business case. M&V also helps to ensure that the project continues to deliver energy savings throughout project life.

An M&V plan needs to be developed early in the project development stage. The plan should include all components—namely, the performance model, the metering plan, and baseline adjustments.

A cooling water system retrofit that supplies water to several molding machines was retrofitted for energy efficiency. The retrofit included installation of VFD on pumps, right sizing pumps, and modification of the piping system to optimize the pumping energy. While working on the M&V plan, only base case pumping power and retrofit case pumping power were considered. It was proposed to normalize the same with the flow through the molding machines. The plant decided to add more molding machines, and an expansion was also required in the cooling plant to handle the cooling load. These changes required baseline adjustment for the increase in the system flow. Unfortunately, this was not part of the original M&V plan, so the plant had to make several assumptions for the adjustment.

System Baseline Measurements

A list of prioritized projects has a benefit in terms of baseline measurement. This is more for the facilities that are in the early stages of their energy program. A floating head pressure control project needs two winters—one in the pre-project stage and one in the post-project stage—to do an M&V. The project plan also needs to consider the same.

Several projects install a submetering system that collects data on a continuous basis. The real question is whether the data are providing any value. For example, in a chilled water plant context, electricity data should be collected with cooling data, like tons.

The installation of the metering at the scoping stage would help not only for the M&V but also in the process of continuous optimization of the system.

The baseline measurements should also document various operating conditions. For example, in a compressed air system project, documentation should include details of compressor operating conditions, such as lead and lag compressors, the pressure setting of the cascade operation, production equipment that is using compressed air, etc. This information becomes useful if there are any changes to plant operating conditions.

Project Implementation

The issues related to project implementation relate to project delays and last-minute changes in the project scope. If the project is on a tight budget, then the efficiency piece of the project might be removed to cut corners. It might be useful to carry out a quick financial analysis to determine how the project would be affected in scope by last-minute changes.

Project implementation should also include documentation and operator training—without which, the project would not deliver on its objectives.

Case Study

The project involves optimization of controls in an industrial office building. Figure 7-9 outlines that in 2010 the building was not operating optimally. Through trial and error, building operators established a more efficient operation protocol in 2011. However, in 2013 the building operations management team changed. Due to a lack of understanding about the operation of the building, the new team operated the building in a safe mode with fewer setbacks and equipment running for longer hours than necessary.

Figure 7-9 presents a CUSUM analysis of the building using 2010 as the base year. A downward slope of the CUSUM graph indicates a reduction in energy and an upward slope indicates an increase in energy consumption. As the figure shows, the building operator was operating the building well in 2011, which resulted in energy savings during that period. In 2013, due to the change in the operator and the building operating in safe mode, monthly consumption rose. The CUSUM analysis shows a slope change around 2013, suggesting an increase in energy consumption The positive slope change around late 2014 shows a reduction in energy consumption addressing some of this operational inefficiency.

Project Commissioning

Often projects are not commissioned properly, which can result in poor post-project performance. A third-party commissioning agent should be included in the project scope. Short-term M&V can determine project savings.

Project Pitfalls 203

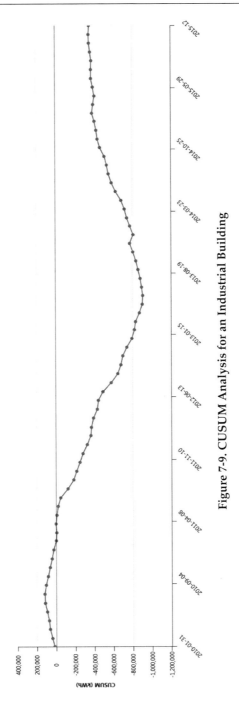

Figure 7-9. CUSUM Analysis for an Industrial Building

Chapter 8

Energy Management Best Practices

THE ENERGY MANAGEMENT CONTINUUM

Different companies take different approaches to energy management. Some pursue a project-based approach, and others incorporate best practices to implement a culture of continuous improvement based on data. These companies are able to reduce energy costs on a continuous basis, but they also are able to capitalize on this in terms of others values, such as brand image and corporate social responsibility.

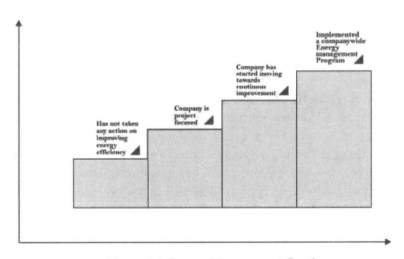

Figure 8-1. Energy Management Continuum

Companies can maximize the value of an energy management program depending on their position in the energy continuum. Companies on the low side of the continuum, for example, might be seen as wasting precious resources trying to save money. These companies focus more on implementing an energy efficiency project without developing the system and organizational structure to support energy management. As a result, they cannot sustain energy reduction.

Successful energy management programs are based on three major components—organizational, behavioral, and technological. Energy management leaders often implement complex measures because they have already built the necessary culture and infrastructure to adopt such technologies. Companies implementing a sustainable program will have sustainable energy reduction and will have more value from the program.

Figure 8-2 shows a company that was aggressive in implementing an energy program in the first few months, but later, they could not maintain the momentum. As a result, the cumulative energy consumption started increasing. In the second case, however, the company implemented a continuous improvement model that led to a consistent reduction of energy.

NECESSARY ELEMENTS FOR THE SUCCESS OF A SUSTAINABLE ENERGY PROGRAM

Senior Management Commitment

Senior management commitment is critical for the success of any energy program. The commitment can be reflected in terms of setting goals, providing resources, and organizational support toward achieving goals. Senior management commitment is reflected in terms of an energy policy. The policy is a written statement of the commitment toward implementing an energy management program and is often signed by a senior executive. Depending on their position on the energy management ladder, companies at level 3 might have active commitment to having no policy. In terms

Energy Management Best Practices 207

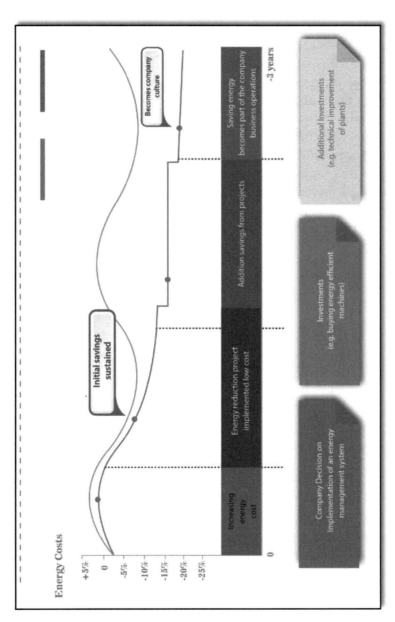

Figure 8-2. Strategic Approach to Energy Management

of organizational support, the companies in level 4 would have the energy management program supported at various levels of management with defined accountability for energy performance; companies at level 2 would have informal organizational responsibility. The companies at level 4 would have financial resources dedicated to supporting the energy management objective, while companies at level 2 only tend to implement the low cost and no cost measures.

Energy Monitoring System

An energy monitoring system provides information on energy consumption for the manufacturing plant. The information can be used together with information on energy drivers, such as production, temperature, etc., to develop a performance matrix.

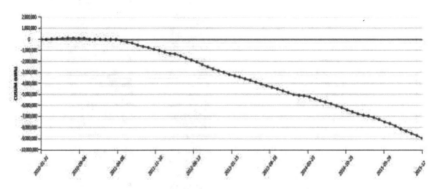

Figure 8-3. CUSUM Analysis (M&T)

Energy monitoring can be done at either the plant level or the equipment level. The other function of an energy monitoring system is to provide data toward identifying energy reduction opportunities and to provide information support toward budgeting. The monitoring system can track operational abnormalities and can help identify corrective action. The monitoring system can be used for commissioning projects relating to energy efficiency. The energy monitoring system can generate reports for various levels of management; based on these reports actions can be taken.

Energy Management Best Practices

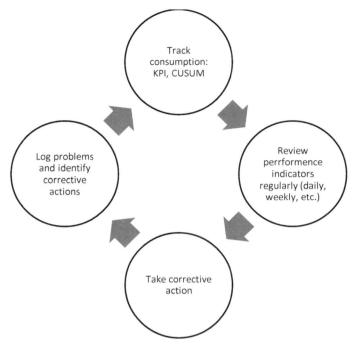

Figure 8-4. Action Oriented Monitoring

For implementing a sustainable energy program, it is important to use the information provided by the energy management information system and take necessary action. This requires management accountability and necessary support.

Employee Engagement

Because employees are important stakeholders in the energy use of a manufacturing plant, employee engagement in the energy management program is a critical factor for success. The employee engagement in the process is reflected in terms of informed decision making regarding energy.

Partnership

The energy management program is also supported by partners outside the company: Utility programs, technology vendors and product innovators, relevant regulation, energy services com-

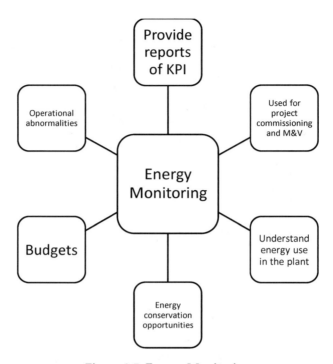

Figure 8-5. Energy Monitoring

panies, industry associations, academic and research institutions, engineering consultants, and others. The partners provide access to technologies, financial incentives, and types of skills. The partners can also help address barriers; the best example in this category is a utility incentive program.

SOME OF THE KEY INITIATIVES TO ADDRESS BARRIERS

Utility Incentive Programs
Embedded Energy Manager and
Program Roving Energy Manager Program

The program was delivered as a part of Ontario's saveONenergy incentive program. The program pays up to 80 percent of the cost of an energy manager for a facility. The energy manager will

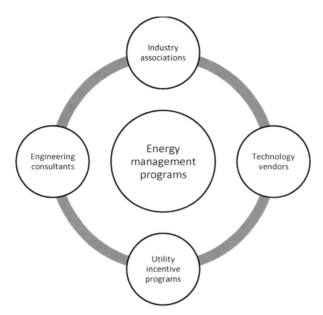

Figure 8-6. Energy Management Partners

have an annual target for energy reduction. The energy manager program essentially helps companies address barriers and helps develop an energy management program in the facility. It also has another program which provides a shared energy manager program for supporting energy management efforts in the facility.

ISO 50001 and Superior Energy Performance

ISO 50001 is an energy management system standard, a framework for industrial and commercial facilities to manage energy. It includes all aspects of energy procurement and use. An energy management system establishes the structure and discipline to implement technical and management strategies that significantly cut energy costs and greenhouse gas emissions and sustain those savings over time. Savings can come from no cost to low cost operational improvements.

Facilities certified to Superior Energy Performance® (SEP™) are leaders in energy management and productivity improvement.

The facilities in SEP have met the ISO 50001 standard and have improved their energy performance up to 30 percent over 3 years.

SEP provides guidance, tools, and protocols to drive deeper, more sustained savings from ISO 50001. To become certified, facilities must implement an energy management system that meets the ISO 50001 standard and demonstrate improved energy performance. An independent third party audits each facility to verify achievements and qualify it at the silver, gold, or platinum level, based on energy performance improvement. This certification emphasizes measureable savings through a transparent process.

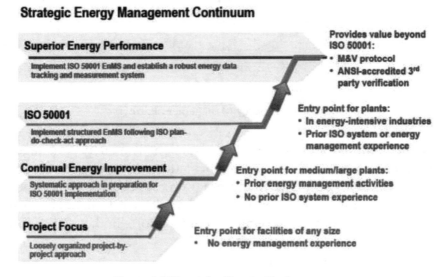

Figure 8-7. Superior Energy Performance

Case Study on Implementing an Energy Management Program
ISO 50001 section 4.4.3 Requires an Energy Review

The organization shall develop, record, and maintain an energy review. The methodology and criteria used to develop the energy review shall be documented. To develop the energy review, the organization shall… omitted sections a), b) and c) are shown below.

The energy review shall be updated at defined intervals, as

well as in response to major changes in facilities, equipment, systems, or processes. [1]

The ISO 50001 energy review is made up of several components and is the key starting point in developing a plan for energy savings. The energy review helps establish the input for the energy baseline and the priority areas for energy improvement projects.

Energy Uses, Consumption and Sources
ISO 50001 section 4.4.3 a):
A) Analyze energy use and consumption based on measurement and other data, i.e.,
- identify current energy sources;
- evaluate past and present energy use and consumption [1].

For clarification, here are several definitions of energy terms:

1. Energy Uses
   a. The manner or the way that energy is used, typically by referring to the type of equipment or process
   b. Examples of energy uses include ventilation fans, air compressors, chill water pumps, high bay lights, convection ovens, conveyor belts, and building heaters.
2. Energy Consumption
   a. The quantity of energy applied in specific energy units
   b. Examples of energy consumption would be in units of kWh or Btu.
3. Energy Sources
   a. Where the energy comes from
   b. For the purposes of ISO 50001 these are typically divided into three areas:
      i. Feedstocks (i.e. process ingredients with energy content)
      ii. Primary energy (i.e., natural gas, coal, diesel fuel, etc.)
      iii. Derived energy (i.e., electricity, steam, solar, wind, biomass, etc.)

A block diagram is useful to identify the basic steps in transforming raw materials into a final product. Understanding these basic transformation steps and then looking at them in terms of required energy input is very helpful in determining where potential process energy savings can be realized.

Following is an example of the process block diagram for a small waste water treatment plant. In this case, the raw material is residential sewage and the final product is clean water.

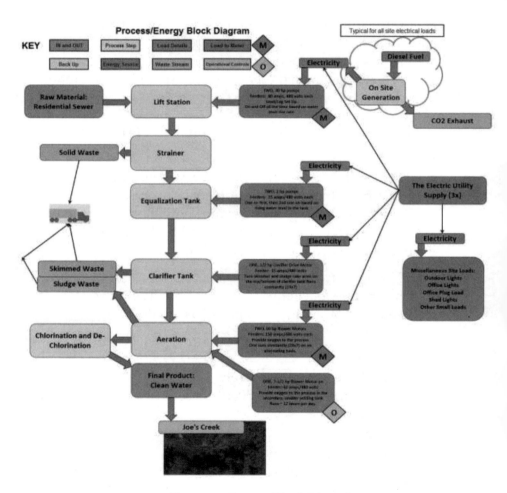

Figure 8-8. Process Block Diagram

Based on the site visit, historical energy bills, and the process block diagram, AE determined the following for the example site:

1. Energy Uses
   a. Electrically driven motors
      i. Two lift station pumps
      ii. Two equalization tank pumps
      iii. One clarifier drive motor
      iv. Two large primary blowers on the main settling tank
      v. One small blower on the smaller settling tank
   b. Shed, office, and outdoor lighting loads and plug loads
   c. Back-up diesel generator
2. Energy Consumption
   a. An energy balance based on actual equipment nameplate data and the historical electric bills revealed the example site consumes an average of:
      i. 529,000 kWh per year electrical energy consumption
      ii. $33,600 per year electrical energy cost
      iii. $0.0635/kWh, electrical unit energy cost
   b. Figure 8-4 shows the electrical energy consumption breakdown by equipment
3. Energy Sources
   a. Electricity
   b. Diesel fuel (not significant under normal operating conditions)

Significant Energy Users

One of the primary reasons for performing an energy review is to find out in detail where the energy is going and what equipment is consuming it. As a part of the energy review, the ISO 50001 standards require the identification of significant energy uses (SEUs):

ISO 50001 section 4.4.3 b):
B) Based on the analysis of energy use and consumption, identify the areas of significant energy use, i.e.
 • identify the facilities, equipment, systems, processes and

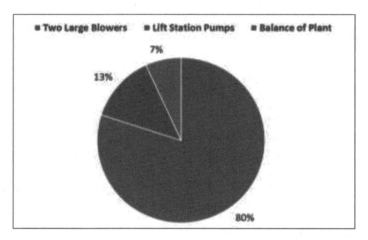

Figure 8-9. Electrical Consumption Splits

personnel working for, or on behalf of, the organization that significantly affect energy use and consumption;
- identify other relevant variables affecting significant energy uses;
- determine the current energy performance of facilities, equipment, systems and processes related to identified significant energy uses;
- estimate future energy use and consumption. [1]

Based on the analysis results shown in Figure 8-4, the SEUs for the example site are:
1. The two large blowers
2. The two lift station pumps

These two items account for approximately 93 percent of the energy consumed at the example site.

ISO 50001 requires the consideration of relevant variables that could impact the SEUs. In general, relevant variables can include things like production quantities, weather parameters, seasonal impacts, hours worked, and raw material characteristics. For SEP, consideration of relevant variables for production level and weather parameters is required.

Typical weather parameters considered are precipitation, cooling degree days or heating degree days, and humidity.

In the case of the example site, production is measured in millions of gallons per day. This is a potential relevant variable for the SEUs.

Looking at weather, the example site operations are outside and the outdoor temperature does not have an impact on energy consumption. Humidity also does not impact the process energy consumption. In some wastewater plants, rain water becomes part of the process flow and could impact the energy consumption. At the example site, rain water does not come into the lift station. Overall, there are no significant impacts from weather on process energy consumption. Therefore, weather is not a relevant variable for the example site SEUs.

On a historical annual average basis back to January 2013, the energy performance for the two example site SEUs is:
1. Large blowers
    a. 420,000 kWh/year
    b. 4,480 kWh/MGAL
2. Lift station pumps
    a. 70,000 kWh/year
    b. 740 kWh/MGAL

Opportunities for Energy Improvement
ISO 50001 section 4.4.3 c):
C) Identify, prioritize and record opportunities for improving energy performance.
NOTE: Opportunities can relate to potential sources of energy, use of renewable energy, or other alternative energy sources, such as waste energy. [1]

Based on the energy balance and the selection of SEUs, the priorities for opportunities for energy improvements at the example site are:
1. Large blowers (80 percent of energy consumption)
2. Lift station pumps (13 percent of energy consumption)

For the lift station pumps, the motors run intermittently and for short periods of time. When they do run, they need to run at full load to push the water up to the strainer. Based on this, a variable frequency drive (VFD) is not practical. Opportunities for energy efficiency on the lift station pumps could include the use of NEMA premium efficiency motors, proper pump and motor maintenance and lubrication, and proper periodic cleaning of the lift station basin.

For the large blowers, one of these two 60 hp blowers runs 24 hours a day, seven days a week, or roughly 4,300 hours per year on each blower. Currently, these blowers are just "on" during plant operations. There are not any existing procedural or engineering controls for these blowers. They just run.

Aeration is needed to help maintain the biological oxygen demand in the wastewater. In the past, the example site blowers did have a dissolved oxygen meter feedback loop to the blowers, but when the blowers were turned off for an extended period of time, sludge settled on top of the aeration piping, blocking the nozzles and causing major problems. This system was eliminated.

Some type of procedural control, motor VFD, or oxygen level feedback would definitely provide energy savings. Anything that can reduce the blower run hours or blower load while running will reduce the blower energy consumption. Other opportunities for energy efficiency improvements on the blowers could include the use of NEMA premium efficiency motors, proper blower and motor maintenance and lubrication, and proper periodic cleaning of the aeration nozzles and settling basins.

In preparation for the implementation of energy improvement projects, installation of submetering is frequently a good idea in order to collect before and after improvement data. Figure 8-10 shows a picture of the WattNode Pulse energy meter installed at the example site.

AE assisted the example site personnel with:

1. Selection of loads to meter
2. Identification of appropriate metering equipment

Energy Management Best Practices

3. Quote review for metering installation
4. Start-up of metering equipment
5. Initiation of data collection
6. Download of metering data

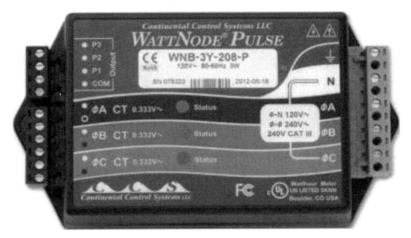

Figure 8-10. WattNode Pulse Metering Device

Energy Baseline

ISO 50001 definitions section 3.13:

Energy performance indicator (EnPI): quantitative value or measure of energy performance, as defined by the organization NOTE EnPIs could be expressed as a simple metric, ratio or a more complex model. [1]

ISO 50001 section 4.4.4:

The organization shall establish an energy baseline(s) using the information in the initial energy review, considering a data period suitable to the organization's energy use and consumption. Changes in energy performance shall be measured against the energy baseline(s).

Adjustments to the baseline(s) shall be made in the case of one or more of the following:
- EnPIs no longer reflect organizational energy use and consumption, or

- there have been major changes to the process, operational patterns, or energy systems, or
- according to a predetermined method.

The energy baseline(s) shall be maintained and recorded. [1]

ISO 50001 requires the establishment of an energy baseline. This baseline is typically a 12-month period corresponding to a calendar year. This is convenient and works well with company recordkeeping. The baseline is established with data collected during the energy review, typically from historical energy bills.

The energy baseline is key because it is the point from which all future energy improvements are measured. Proper selection of the energy baseline time period and the data is very important. Data outliers must be identified and explained.

Figure 8-11 shows historical electrical energy usage for the example site, from which a baseline of energy consumption would need to be selected.

The energy data baseline, along with production data baseline and any other relevant variables, will be combined to form the overall energy performance baseline.

Energy Performance Indicators
ISO 50001 section 4.4.5

The organization shall identify EnPIs appropriate for monitoring and measuring its energy performance. The methodology for determining and updating the EnPIs shall be recorded and regularly reviewed.

EnPIs shall be reviewed and compared to the energy baseline as appropriate. [1]

A good EnPI for the example site could be kWh/MGAL (kilowatt hours of electrical energy consumption per million gallons of wastewater processed). Figure 8-12 shows an example of a possible EnPI trend graph.

The whole point of the ISO 50001 program is to have a consistent and ongoing improvement trend in EnPI performance.

Energy Management Best Practices 221

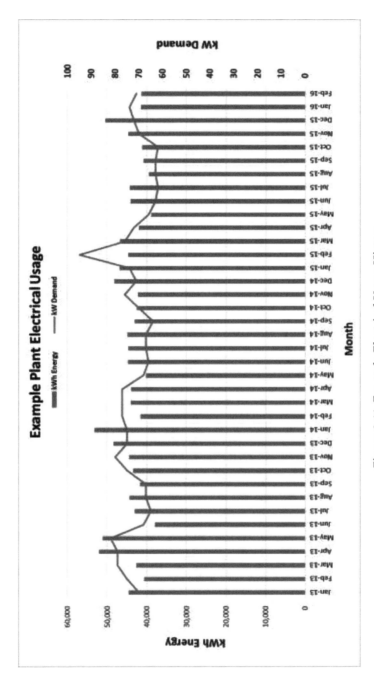

Figure 8-11. Example Electrical Usage History

Energy Objectives, Targets, and Action Plans
ISO 50001 section 4.4.6

The organization shall establish, implement and maintain documented energy objectives and targets at the relevant functions, levels, processes or facilities within the organization. Time frames shall be established for achievement of the objectives and targets.

The objectives and targets shall be consistent with the energy policy. Targets shall be consistent with the objectives.

When establishing and reviewing objectives and targets, the organization shall take into account legal requirements and other requirements, significant energy uses and opportunities to improve energy performance, as identified in the energy review. It shall also consider its financial, operational and business conditions, technological options and the views of interested parties.

The organization shall establish, implement and maintain action plans for achieving its objectives and targets.

The action plans shall include:
- designation of responsibility;
- the means and time frame by which individual targets are to be achieved;
- a statement of the method by which an improvement in energy performance shall be verified;
- a statement of the method of verifying the results.

The action plans shall be documented, and updated at defined intervals. [1]

Understanding the difference between an objective, a target, and an action plan can be confusing. One way to think about it is as a hierarchy of achievement.

The organization sets its scope and boundaries and writes its own specific energy policy. Objectives are more strategic in nature and are set up to meet the requirements of the energy policy. Targets are more tactical in nature and are set up to help meet the objectives. The action plan is the actual detailed procedures at the execution level that define who, what, when, where, why and how

Energy Management Best Practices 223

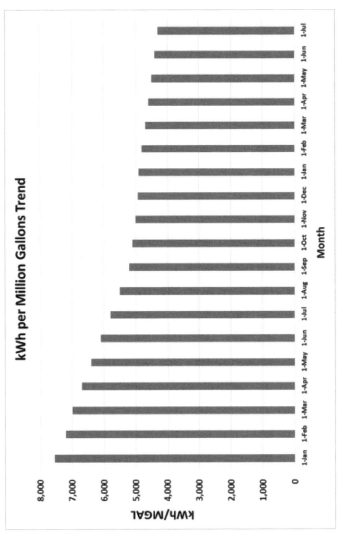

Figure 8-12. Example of an EnPI Trend Graph

for achieving a specific target. Figure 8-13 shows a graphic that represents this hierarchy of achievement.

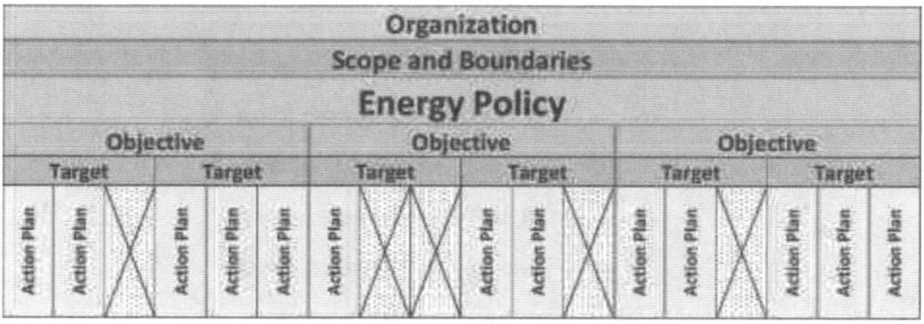

Figure 8-13. Hierarchy of Achievement for ISO 50001

ENERGY MANAGEMENT REVIEW CONTOURS

The energy management contour can be developed based on organizational achievement on different energy management best practices, such as energy management policy, metering, capacity building, etc. In that way, it is a self-assessment tool. The energy management contours can be used to develop an understanding about the state of energy management program in a facility. The energy management best practices are ranked on a scale from 0 to 10; 0 corresponds to almost no action in a best practice area while 10 corresponds to a fully implemented best practice.

The energy management contour can also point toward the kinds of strategies and tactics that can move an energy program to the next level. It is important to understand that all best practices are interrelated. For example, if a company has a high score on an energy policy goal and investment it is likely the company is pursuing a more project-based approach in its energy program. Similarly, a company having a higher score in energy policy, energy management information system, and organizational structure is likely to have sustainable energy reduction.

Energy Management Best Practices 225

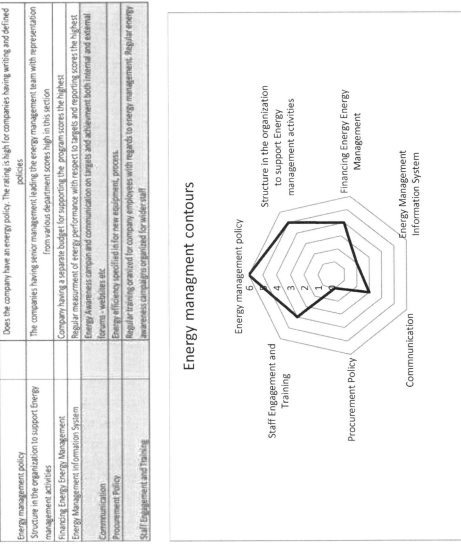

Energy management policy	Does the company have an energy policy. The rating is high for companies having writing and defined policies
Structure in the organization to support Energy management activities	The companies having senior management leading the energy management team with representation from various department scores high in this section
Financing Energy Energy Management	Company having a separate budget for supporting the program scores the highest
Energy Management Information System	Regular measurement of energy performance with respect to targets and reporting scores the highest
Communication	Energy Awareness campain and communication on targets and achievement both internal and external forums - websites etc
Procurement Policy	Energy efficiency specified in for new equipment, process.
Staff Engagement and Training	Regular training oranized for company employees with regards to energy management. Regular energy awareness campaigns organized for wider staff

Figure 8-14. Energy Management Contours

Common Barriers

- Senior management might have the perspective that energy is a fixed cost and really nothing can be done on it.
- Companies are trying to implement an energy management program, but they don't have access to technology and best practices.
- Small and mid-sized companies struggling to survive have limited resources to implement an energy program.
- Companies are being restructured and some of their plants will be relocated, and thus, they would not like to invest in an energy program.
- Companies see the value of energy management but are pursuing a project-based approach to saving energy.

Strategies to Address Barriers

Different barriers can prohibit improving an energy management program, and these can range from weak organizational commitment to lack of resources. Tactics to address some barriers are given below.

Weak Management Commitment

The value of energy management programs is not always seen by senior management. The energy champion's role here is to identify low cost and no cost measures, such as turning off idling equipment. The results of these initiatives should be visible to senior management. A simplified method of tracking savings might be another good idea to demonstrate the results to senior management. In addition, there might be opportunities to add some energy efficiency improvement to capital projects, like replacing chillers or compressors. Again, a measurement and verification plan would help show senior management some of the results of the initiatives. Utility programs often offer incentives for energy efficiency projects; these would help generate resources for some of the energy efficiency projects. Communicating the results is extremely important in demonstrating the value of these energy management initiatives.

Lack of Resource to Support Energy Management Initiatives

In many cases, efficiency projects at industrial facilities face the challenge of having to navigate an investment decision-making framework that places heavy emphasis on optimizing manufacturing processes and ensuring continuous operation of plant assets. Some tactics to address this include taking advantage of utility incentive programs to bring down the payback. Financial analysis beyond simple payback to show management the value the project would bring over the full life of the project. A detailed system approach can reveal additional opportunities for savings; for example, in a compressor air upgrade project, the reduction of leakages can reduce the size of the compressor capacity and the cost of the system and also increase energy savings.

Access to Technology and Resources

Some companies have an energy management program but do not have access to skills and technology to improve the program. The internet can provide access to some of the skills and tools required in this area. Programs such as RETScreen Expert from Natural Resources Canada can help in addressing some of these barriers. These issues are usually more prevalent in developing countries. Organizations such as the Association of Energy Engineers provide certification programs and other relevant courses in the area of energy management. Some courses are available online.

Companies Pursuing a Project-based Approach

Companies pursuing a project-based approach would benefit from having a CUSUM analysis done on the utility bills, showing the results of a project-based approach. These barriers might be addressed by providing training on the value of a monitoring and targeting system. Surely there can be other barriers that might be responsible for this kind of approach, such as a lack of dedicated resources for energy management.

IMPROVING OR IMPLEMENTING AN ENERGY MANAGEMENT PROGRAM

Companies can be at different stages on the energy management ladder based on whether they have implemented best practices. Improving an energy management program to the next level or even establishing a program involves management engagement; this may be as simple as support to implement low cost measures.

The next level is understanding where the company stands in terms of the energy management program, business issues, expansion plans, strategic goals, regulatory compliance, operational challenges, and how energy is being used in the facility, including a list of energy reduction opportunities.

Based on an understanding of the energy and related issues of the facility, it is important to establish a baseline and develop a list of performance indicators—a list of energy reduction opportunities. Clearly, an energy management program can be improved by working on it regularly.

References
Stowe, Michael; Haggis, Staci (2016). ISO 50001/SEP and strategic energy management, Advanced Energy.

Paul Scheihing Technology Manager US Department of Energy, Accelerated Energy Savings through the Superior Energy Performance Program presented at the World Engineering Congress Conference, Washington DC 2016

https://www.carbontrust.com/resources/tools/energy-management-self-assessment-tool/

Chapter 9

Application of Energy Monitoring and Targeting for Industrial Plants

Energy monitoring is a critical component to an energy management program. It provides useful information on the facility's energy usage, identifies improvement opportunities, helps to set energy reduction goals, and monitors energy performance improvement.

COMPONENTS OF AN ENERGY MANAGEMENT INFORMATION SYSTEM

The components of any energy management information system (EMIS) involve the following data acquisition: Installation of submeters, sensors, and transducers to collect data that have bearing on energy consumption.

Data analysis is carried out on the collected data, which would include CUSUM analysis, generation of trends, interval data analysis, and development of KPI. The data analysis helps in measuring energy performance, setting targets, and reviewing energy performance against the target on a continuous basis.

A reporting component would help in generating reports for different levels of management.

Specifically, the EMIS would help in the following tasks:

1. detecting any energy performance deviation based on CUSUM analysis;

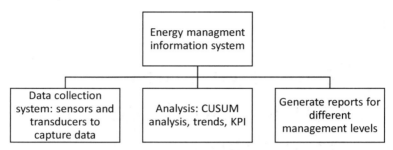

Figure 9-1. Energy Management Information System

2. generating key performance indicators;
3. providing energy performance information for an implemented project;
4. carrying out energy optimization in the process;
5. monitoring key process parameters and setting alarms if the parameters are above a critical threshold;
6. providing useful information on energy budgets; and
7. helping to develop standard operating practices.

The energy monitoring process involves setting up energy accounting centers or cost centers. The cost centers can be based on energy use in the facility or they can be based on process operation.

CUSUM ANALYSIS

The following section discusses the concepts and application of a CUSUM analysis, which is an important part of monitoring and targeting.

Case Study 1

A manufacturing facility with a three-shift operation implemented a lighting project in which it replaced metal halide lamps with LED lamps in one section of the plant; this resulted in saving

100 kW. The project was completed in May 2014. At the end of nine months, the plant was happy with the retrofit and decided to perform a lighting retrofit in another section of the plant to add further savings of 100 kW.

The manufacturing plant decided to optimize the chilled water plant operation, like chiller sequencing and installing VFD. The total savings resulting from the optimization of the plant was about 300 kW.

The plant manager wants a report showing the savings for 24 months. The monthly production level has not changed throughout this entire period.

As clearly noticed in Figure 9-2 at points A, B, and C, there is a change in the slope showing change in the energy savings. The savings increases at points A, B, and C, but the savings goes back to zero at point D.

The important takeaways from the analysis are as follows:

- There is a change in slope whenever there is a change in energy savings.

- A positive change indicates a reduction in energy savings and a negative change in the slope indicates an increase in energy savings.

Case Study 2

A plastic extrusion facility implemented a compressed system upgrade and the energy manager wanted to track improvement using a CUSUM analysis, and to scatter plot energy and production to create the regression model. In this example, the baseline period is selected for first 12 months.

A scatterplot of the independent variable (production) and dependent variable (energy) was created using MS Excel. The performance model for it was established by right clicking the mouse in the scatterplot and selecting the format trend line and selecting the display equation on chart and display R2 on chart. The value of R2 shows that the baseline equation is a good fit, as the value of R2 is greater than 0.75.

Table 9-1. CUSUM Analysis, Case 1

Month	Demand Savings (KW)	Energy Savings (KWh)	CUSUM
MONTH1	100	72000	-72000
MONTH2	100	72000	-144000
MONTH3	100	72000	-216000
MONTH4	100	72000	-288000
MONTH5	100	72000	-360000
MONTH6	100	72000	-432000
MONTH7	100	72000	-504000
MONTH8	100	72000	-576000
MONTH9	100	72000	-648000
MONTH10	200	144000	-792000
MONTH11	200	144000	-936000
MONTH12	200	144000	-1080000
MONTH13	200	144000	-1224000
MONTH14	500	360000	-1584000
MONTH15	500	360000	-1944000
MONTH16	500	360000	-2304000
MONTH17	500	360000	-2664000
MONTH18	500	360000	-3024000
MONTH19	0	0	-3024000
MONTH20	0	0	-3024000
MONTH21	0	0	-3024000
MONTH22	0	0	-3024000
MONTH23	0	0	-3024000
MONTH24	0	0	-3024000

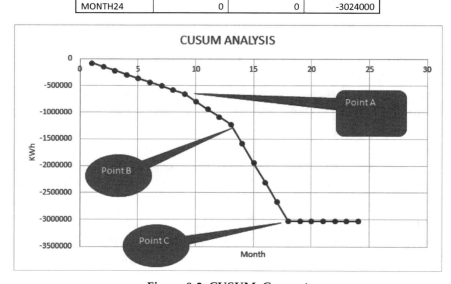

Figure 9-2. CUSUM, Concept

Application of Energy Monitoring and Targeting for Industrial Plants

Table 9-2. Raw Data

Month	Production (lbs.)	kWh
1	3750	562904
2	2000	412892
3	1500	363056
4	1250	350268
5	4250	586400
6	4500	619092
7	3000	486300
8	1000	325744
9	2750	462344
10	2250	423636
11	1000	335664
12	1250	345088
13	3500	503568
14	3875	555864
15	4125	559688
16	4750	609096
17	1000	311152
18	1375	330844
19	3750	497268
20	2000	378708
21	1575	338512
22	2750	432164
23	1750	356460
24	4250	545552
25	4750	565712
26	4000	564860
27	3000	473276
28	4750	610024
29	2000	397068
30	2250	377872
31	4500	560752
32	1750	365048
33	1250	312992
34	3875	512020
35	4175	524012
36	3000	436768

The performance model is then used to predict the energy consumption for the entire dataset. This means that if the plant were operated in baseline conditions, the energy consumption would have been the predicted value based on the baseline performance model. The difference in the actual energy consumption and the predicted energy consumption would give the value of energy savings. The cumulative energy savings for each month is also shown.

RETSCREEN EXPERT

RETScreen Expert is a clean energy management software for energy efficiency, cogeneration and renewable energy projects. The software was developed by Natural Resources Canada. The energy performance tab has the capability to conduct CUSUM analysis and measurement and verification for a dataset. Depending

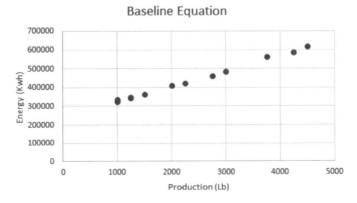

Figure 9-3. Scatterplot

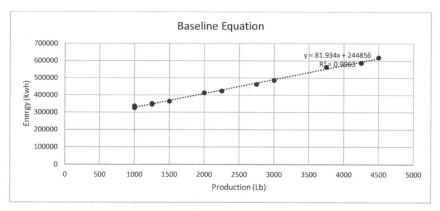

Figure 9-4. Baseline Model

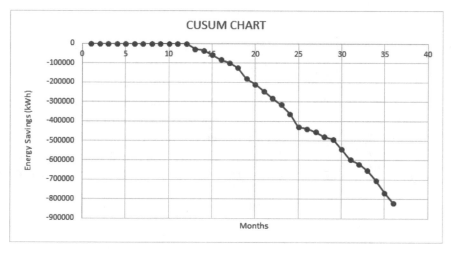

Figure 9-5. CUSUM Analysis

on the independent variables used, the CUSUM analysis process involves four major steps: Weather station selection, data preparation, analytics, and reporting.

Daily weather data from NASA can be downloaded in the tool. Various weather stations across the globe can be selected

Application of Energy Monitoring and Targeting for Industrial Plants 235

Month	Production (Lb)	kWh	Predicted Energy (KWh)	Savings (Kwh) [Actual-Predicted]	CUSUM
1	3750	562904	552109		0
2	2000	412892	408724		0
3	1500	363056	367757		0
4	1250	350268	347274		0
5	4250	586400	593076		0
6	4500	619092	613559		0
7	3000	486300	490658		0
8	1000	325744	326790		0
9	2750	462344	470175		0
10	2250	423636	429208		0
11	1000	335664	326790		0
12	1250	345088	347274		0
13	3500	503568	531625	-28057	-28057
14	3875	555864	562350	-6486	-34543
15	4125	559688	582834	-23146	-57689
16	4750	609096	634043	-24947	-82636
17	1000	311152	326790	-15638	-98274
18	1375	330844	357515	-26671	-124945
19	3750	497268	552109	-54841	-179785
20	2000	378708	408724	-30016	-209801
21	1575	338512	373902	-35390	-245191
22	2750	432164	470175	-38011	-283202
23	1750	356460	388241	-31781	-314982
24	4250	545552	593076	-47524	-362506
25	4750	565712	634043	-68331	-430836
26	4000	564860	572592	-7732	-438568
27	3000	473276	490658	-17382	-455950
28	4750	610024	634043	-24019	-479969
29	2000	397068	408724	-11656	-491625
30	2250	377872	429208	-51336	-542960
31	4500	560752	613559	-52807	-595767
32	1750	365048	388241	-23193	-618960
33	1250	312992	347274	-34282	-653241

from the weather map.

Data preparation essentially involves entering the data in a specified format in Excel. Separate files can be entered in the tool manually or imported as an Excel file. Data preparation can also involve merging separate files to get the data for energy consumption and related drivers in one table.

Once the data preparation is complete, regression analysis is performed on the dataset. This tool has the capability to perform multivariate regression analysis and to select different types of baseline equations, like linear, polynomial, etc. The tool also provides capability to remove outliers and to vary the baseline period. Once the baselines are finalized, the CUSUM and measurement and verification (M&V) for selected baseline can be carried out.

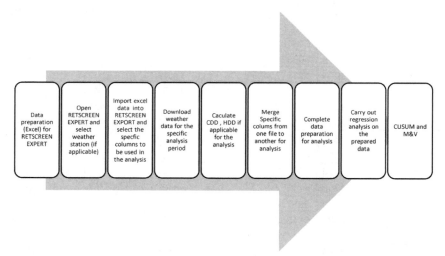

Figure 9-6. RETScreen Expert Analysis Stages

Example 1

The example discussed here is taken from a manufacturing facility that has implemented some operational measures. Monthly electricity consumption from utility meters and production for the corresponding period was compiled in a spreadsheet.

The data in the above table are organized in a format recog-

Application of Energy Monitoring and Targeting for Industrial Plants 237

nized by RETScreen Expert. RETScreen Expert is launched and the data tab is opened because no weather variable is used. In the data tab, the Excel file is imported into RETScreen Expert by clicking the user defined section.

Table 9-3. Data Preparation for RETScreen Expert

DATE	Productio	Electricity (KWh)	YEAR	MONTH	DAY	Productio	Electricity
2011-02-28	21883724	1003671	2011	2	28	21883724	1003671
2011-03-30	30516208	1266471	2011	3	30	30516208	1266471
2011-04-30	22402064	1130922	2011	4	30	22402064	1130922
2011-05-30	26377040	1239633	2011	5	30	26377040	1239633
2011-06-30	34621868	1253925	2011	6	30	34621868	1253925
2011-07-30	18837136	1128066	2011	7	30	18837136	1128066
2011-08-30	25643440	1122408	2011	8	30	25643440	1122408
2011-09-30	10906000	824100	2011	9	30	10906000	824100
2011-10-30	6642072	815424	2011	10	30	6642072	815424
2011-11-30	14211756	932208	2011	11	30	14211756	932208
2011-12-30	7943840	781131	2011	12	30	7943840	781131
2012-01-30	19856988	1063749	2012	1	30	19856988	1063749
2012-02-29	25757168	1049199	2012	2	29	25757168	1049199
2012-03-30	24989736	1109256	2012	3	30	24989736	1109256
2012-04-30	19674308	985605	2012	4	30	19674308	985605
2012-05-30	22686016	1105347	2012	5	30	22686016	1105347
2012-06-30	19316556	1144950	2012	6	30	19316556	1144950
2012-07-30	13814620	1105368	2012	7	30	13814620	1105368
2012-08-30	15811080	869274	2012	8	30	15811080	869274
2012-09-30	11519780	736875	2012	9	30	11519780	736875
2012-10-30	15048584	806268	2012	10	30	15048584	806268
2012-11-30	14315464	850947	2012	11	30	14315464	850947
2012-12-30	5857676	694986	2012	12	30	5857676	694986

Table 9-3B.

DATE	kWh	Production (kG)
31-Jan-09	1,249,515	4042004
28-Feb-09	1,262,706	3678599
28-Mar-09	1,443,522	4392482
25-Apr-09	1,588,810	3702505
23-May-09	1,452,679	3698677
20-Jun-09	1,824,707	4284257
18-Jul-09	1,676,455	4196475
15-Aug-09	1,605,593	4032677
12-Sep-09	1,785,546	4058356
10-Oct-09	1,553,378	4422985
7-Nov-09	1,607,458	4356070
5-Dec-09	1,327,738	4290424
2-Jan-10	1,429,442	4050733
30-Jan-10	1,266,127	3948353
27-Feb-10	1,334,264	4045437
27-Mar-10	1,464,800	4428643
24-Apr-10	1,387,411	3637207
22-May-10	1,345,453	3854110
19-Jun-10	1,855,476	4383848
17-Jul-10	1,840,558	4267233
14-Aug-10	1,926,338	4657453
11-Sep-10	1,788,343	4419307
9-Oct-10	1,639,159	4178799
6-Nov-10	1,491,840	4197078
4-Dec-10	1,514,218	4422999
1-Jan-11	1,356,642	4113245
29-Jan-11	1,312,819	3855031
26-Feb-11	1,318,414	3905236
26-Mar-11	1,386,479	3627877
23-Apr-11	1,468,530	4293737
21-May-11	1,532,866	3977644
18-Jun-11	1,717,481	3958815
16-Jul-11	1,862,003	4276656
13-Aug-11	1,920,744	4193334
10-Sep-11	1,750,115	3984125
8-Oct-11	1,738,926	4422765
5-Nov-11	1,489,043	4008228
3-Dec-11	1,470,395	4306055
31-Dec-11	1,447,085	4387293
28-Jan-12	1,344,521	3759330
25-Feb-12	1,347,318	3685234
24-Mar-12	1,503,961	3996821
21-Apr-12	1,509,556	4122037
19-May-12	1,589,741	4188225
16-Jun-12	1,756,642	4473477
14-Jul-12	1,943,122	4485694

Table 9-3B (*Continued*).

DATE	kWh	Production (kG)
11-Aug-12	1,652,213	4289149
8-Sep-12	1,841,490	4050979
6-Oct-12	1,628,903	4403276
3-Nov-12	1,375,290	3609985
1-Dec-12	1,436,828	4046661
29-Dec-12	1,366,898	4000089
26-Jan-13	849789	3114239
23-Feb-13	835430	3081776
23-Mar-13	954218	3953345
20-Apr-13	896782	3036357
18-May-13	712727	3568005
15-Jun-13	1086712	3761762
13-Jul-13	1155896	3924825
10-Aug-13	1146106	3715202
7-Sep-13	1091934	3411551
5-Oct-13	1125873	4108086
2-Nov-13	987505	3594120
30-Nov-13	794964	2793485
28-Dec-13	890908	3860457

The Excel table has several columns, and the relevant columns are selected: YEAR, MONTH, DAY, and USERDEFINED, and match the drop-down names with the column title. The columns not being used in the analysis are ignored by selecting DO NOT IMPORT; once this is complete, "OK" is clicked to move to the next tab.

After the data are imported, the regression tab is clicked on the analytic section. The regression tab allows the system to perform regression with different combinations of variables and time periods. For regression analysis, the independent and dependent variable, which is production and energy in the current example, are selected. The baseline period for the analysis is one year starting January 2011; once the year is adjusted to 1, the option is available to select months as well as days. The regression coefficient moves up to 0.8, and the baseline model is reached. Once the CUSUM tab is clicked for the CUSUM chart, the tool also can generate an M&V report by clicking M&V.

240   Industrial Energy Management Strategies

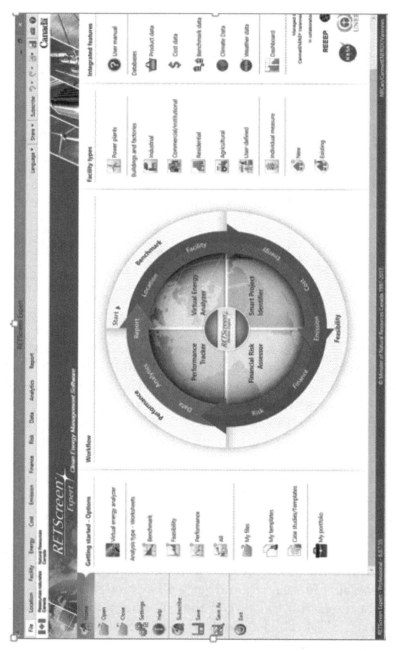

Figure 9-7. RETScreen Expert

Application of Energy Monitoring and Targeting for Industrial Plants 241

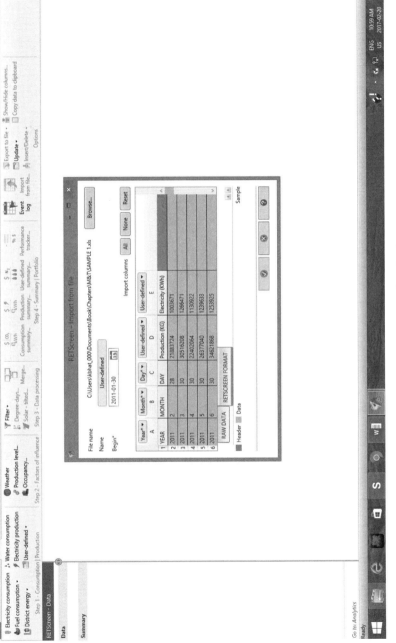

Figure 9-8. User Defined Tab

242 Industrial Energy Management Strategies

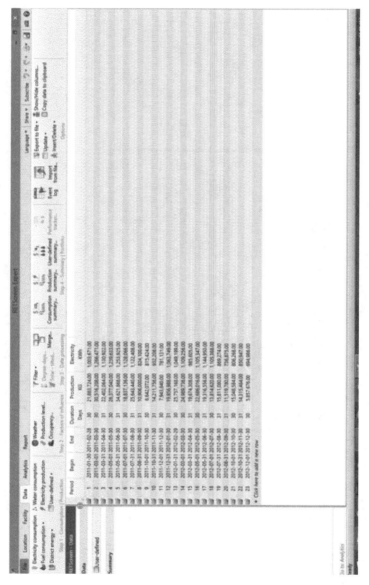

Figure 9-9. Importing Data from Excel

Application of Energy Monitoring and Targeting for Industrial Plants 243

Figure 9-10. Selection Variables for Regression

244   Industrial Energy Management Strategies

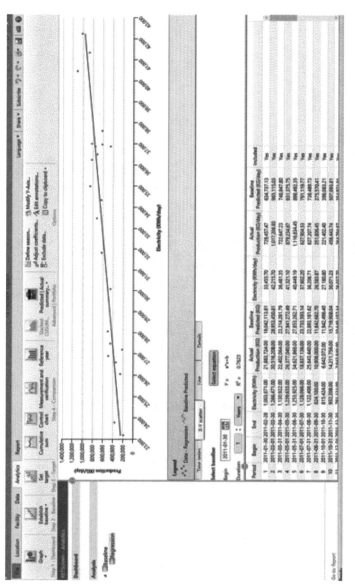

Figure 9-11. Regression Model

Application of Energy Monitoring and Targeting for Industrial Plants 245

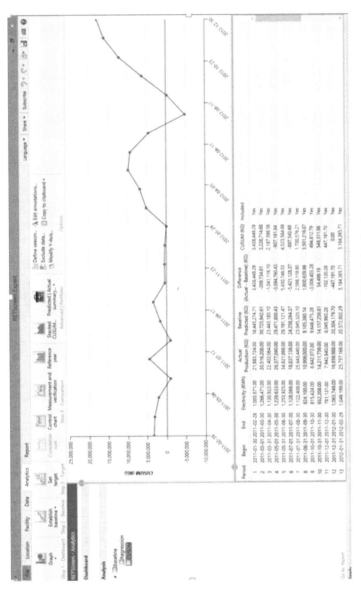

Figure 9-12. CUSUM Analysis

Example 2

The second example uses RETScreen Expert to conduct CU-SUM analysis. Data for production and energy are entered in an Excel spreadsheet.

Table 9-4. Raw Data

DATE	YEAR	MONTH	DAY	kWh	Production
31-Jan-09	2009	1	31	1,249,515	4042004
28-Feb-09	2009	2	28	1,262,706	3678599
28-Mar-09	2009	3	28	1,443,522	4392482
25-Apr-09	2009	4	25	1,588,810	3702505
23-May-09	2009	5	23	1,452,679	3698677
20-Jun-09	2009	6	20	1,824,707	4284257
18-Jul-09	2009	7	18	1,676,455	4196475
15-Aug-09	2009	8	15	1,605,593	4032677
12-Sep-09	2009	9	12	1,785,546	4058356
10-Oct-09	2009	10	10	1,553,378	4422985
7-Nov-09	2009	11	7	1,607,458	4356070
5-Dec-09	2009	12	5	1,327,738	4290424
2-Jan-10	2010	1	2	1,429,442	4050733
30-Jan-10	2010	1	30	1,266,127	3948353
27-Feb-10	2010	2	27	1,334,264	4045437
27-Mar-10	2010	3	27	1,464,800	4428643
24-Apr-10	2010	4	24	1,387,411	3637207
22-May-10	2010	5	22	1,345,453	3854110
19-Jun-10	2010	6	19	1,855,476	4383848
17-Jul-10	2010	7	17	1,840,558	4267233
14-Aug-10	2010	8	14	1,926,338	4657453
11-Sep-10	2010	9	11	1,788,343	4419307
9-Oct-10	2010	10	9	1,639,159	4178799
6-Nov-10	2010	11	6	1,491,840	4197078
4-Dec-10	2010	12	4	1,514,218	4422999
1-Jan-11	2011	1	1	1,356,642	4113245
29-Jan-11	2011	1	29	1,312,819	3855031
26-Feb-11	2011	2	26	1,318,414	3905236
26-Mar-11	2011	3	26	1,386,479	3627877
23-Apr-11	2011	4	23	1,468,530	4293737
21-May-11	2011	5	21	1,532,866	3977644
18-Jun-11	2011	6	18	1,717,481	3958815
16-Jul-11	2011	7	16	1,862,003	4276656
13-Aug-11	2011	8	13	1,920,744	4193334
10-Sep-11	2011	9	10	1,750,115	3984125
8-Oct-11	2011	10	8	1,738,926	4422765
5-Nov-11	2011	11	5	1,489,043	4008228

Application of Energy Monitoring and Targeting for Industrial Plants 247

Table 9-4 (*Continued*). Raw Data

DATE	YEAR	MONTH	DAY	kWh	Production
3-Dec-11	2011	12	3	1,470,395	4306055
31-Dec-11	2011	12	31	1,447,085	4387293
28-Jan-12	2012	1	28	1,344,521	3759330
25-Feb-12	2012	2	25	1,347,318	3685234
24-Mar-12	2012	3	24	1,503,961	3996821
21-Apr-12	2012	4	21	1,509,556	4122037
19-May-12	2012	5	19	1,589,741	4188225
16-Jun-12	2012	6	16	1,756,642	4473477
14-Jul-12	2012	7	14	1,943,122	4485694
11-Aug-12	2012	8	11	1,652,213	4289149
8-Sep-12	2012	9	8	1,841,490	4050979
6-Oct-12	2012	10	6	1,628,903	4403276
3-Nov-12	2012	11	3	1,375,290	3609985
1-Dec-12	2012	12	1	1,436,828	4046661
29-Dec-12	2012	12	29	1,366,898	4000089
26-Jan-13	2013	1	26	849789	3114239
23-Feb-13	2013	2	23	835430	3081776
23-Mar-13	2013	3	23	954218	3953345
20-Apr-13	2013	4	20	896782	3036357
18-May-13	2013	5	18	712727	3568005
15-Jun-13	2013	6	15	1086712	3761762
13-Jul-13	2013	7	13	1155896	3924825
10-Aug-13	2013	8	10	1146106	3715202
7-Sep-13	2013	9	7	1091934	3411551
5-Oct-13	2013	10	5	1125873	4108086
2-Nov-13	2013	11	2	987505	3594120
30-Nov-13	2013	11	30	794964	2793485
28-Dec-13	2013	12	28	890908	3860457

RETScreen reads data in a specific format; therefore, the data are then converted in the following format using Excel function such as year (date), month (date), and day (date). The created Excel file is saved on the desktop or another location.

Example 3

RETScreen Expert is opened, and the performance tab is selected under the "getting started" option. In the top row is a tab named "location," which allows the user to select a weather station from NASA. The weather station provides weather information for various locations in the world. In the present example, the Toronto International Airport weather station is selected.

Table 9-5. Data Convert to Data Preparation Format

DATE	YEAR	MONTH	DAY	kWh	Production
31-Jan-09	2009	1	31	1,249,515	4042004
28-Feb-09	2009	2	28	1,262,706	3678599
28-Mar-09	2009	3	28	1,443,522	4392482
25-Apr-09	2009	4	25	1,588,810	3702505
23-May-09	2009	5	23	1,452,679	3698677
20-Jun-09	2009	6	20	1,824,707	4284257
18-Jul-09	2009	7	18	1,676,455	4196475
15-Aug-09	2009	8	15	1,605,593	4032677
12-Sep-09	2009	9	12	1,785,546	4058356
10-Oct-09	2009	10	10	1,553,378	4422985
7-Nov-09	2009	11	7	1,607,458	4356070
5-Dec-09	2009	12	5	1,327,738	4290424
2-Jan-10	2010	1	2	1,429,442	4050733
30-Jan-10	2010	1	30	1,266,127	3948353
27-Feb-10	2010	2	27	1,334,264	4045437
27-Mar-10	2010	3	27	1,464,800	4428643
24-Apr-10	2010	4	24	1,387,411	3637207
22-May-10	2010	5	22	1,345,453	3854110
19-Jun-10	2010	6	19	1,855,476	4383848
17-Jul-10	2010	7	17	1,840,558	4267233
14-Aug-10	2010	8	14	1,926,338	4657453
11-Sep-10	2010	9	11	1,788,343	4419307
9-Oct-10	2010	10	9	1,639,159	4178799
6-Nov-10	2010	11	6	1,491,840	4197078
4-Dec-10	2010	12	4	1,514,218	4422999
1-Jan-11	2011	1	1	1,356,642	4113245
29-Jan-11	2011	1	29	1,312,819	3855031
26-Feb-11	2011	2	26	1,318,414	3905236
26-Mar-11	2011	3	26	1,386,479	3627877
23-Apr-11	2011	4	23	1,468,530	4293737
21-May-11	2011	5	21	1,532,866	3977644
18-Jun-11	2011	6	18	1,717,481	3958815
16-Jul-11	2011	7	16	1,862,003	4276656
13-Aug-11	2011	8	13	1,920,744	4193334
10-Sep-11	2011	9	10	1,750,115	3984125
8-Oct-11	2011	10	8	1,738,926	4422765
5-Nov-11	2011	11	5	1,489,043	4008228
3-Dec-11	2011	12	3	1,470,395	4306055
31-Dec-11	2011	12	31	1,447,085	4387293
28-Jan-12	2012	1	28	1,344,521	3759330
25-Feb-12	2012	2	25	1,347,318	3685234
24-Mar-12	2012	3	24	1,503,961	3996821

Table 9-5 (*Continued*). Data Convert to Data Preparation Format

DATE	YEAR	MONTH	DAY	kWh	Production
21-Apr-12	2012	4	21	1,509,556	4122037
19-May-12	2012	5	19	1,589,741	4188225
16-Jun-12	2012	6	16	1,756,642	4473477
14-Jul-12	2012	7	14	1,943,122	4485694
11-Aug-12	2012	8	11	1,652,213	4289149
8-Sep-12	2012	9	8	1,841,490	4050979
6-Oct-12	2012	10	6	1,628,903	4403276
3-Nov-12	2012	11	3	1,375,290	3609985
1-Dec-12	2012	12	1	1,436,828	4046661
29-Dec-12	2012	12	29	1,366,898	4000089
26-Jan-13	2013	1	26	849789	3114239
23-Feb-13	2013	2	23	835430	3081776
23-Mar-13	2013	3	23	954218	3953345
20-Apr-13	2013	4	20	896782	3036357
18-May-13	2013	5	18	712727	3568005
15-Jun-13	2013	6	15	1086712	3761762
13-Jul-13	2013	7	13	1155896	3924825
10-Aug-13	2013	8	10	1146106	3715202
7-Sep-13	2013	9	7	1091934	3411551
5-Oct-13	2013	10	5	1125873	4108086
2-Nov-13	2013	11	2	987505	3594120
30-Nov-13	2013	11	30	794964	2793485
28-Dec-13	2013	12	28	890908	3860457

The next tab is data. The Excel template containing data on production and electricity can be imported into the program using the "user defined" tab. The imported table must define year, month, and day using the drop-down menu for each column. The imported Excel table can have several columns, but only the ones to be used in the analysis are selected, then "do not import" is selected for the rest.

Importing the production and electricity data is complete. The next step is clicking on the weather tab and downloading the data from NASA for the period for which the production and electricity data are available.

The two separate data files are created: One is the NASA weather file and the other is user-defined. Because the analysis can only be carried out with data from one file, the "average air temperature" column is merged to user-defined.

The data presentation is now complete, and analytics is next.

250   Industrial Energy Management Strategies

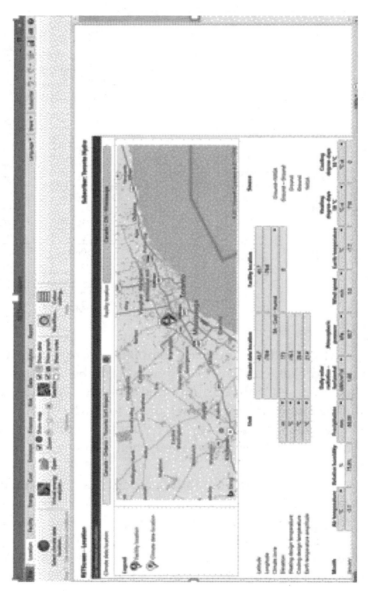

Figure 9-13. Selection of Weather Station

Application of Energy Monitoring and Targeting for Industrial Plants 251

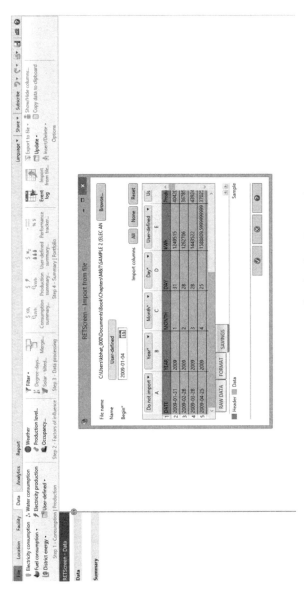

Figure 9-14. User Defined Tab to Import data

252   Industrial Energy Management Strategies

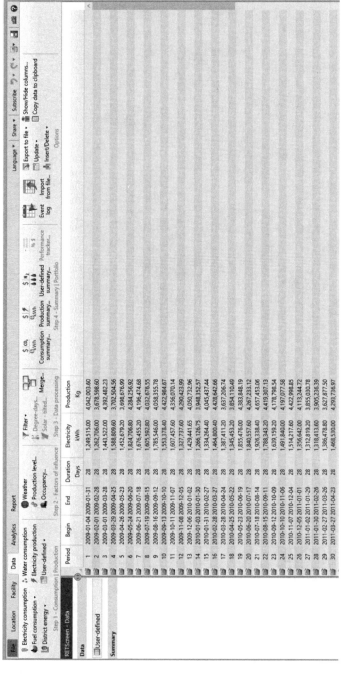

Figure 9-15. Data Imported from Excel

Application of Energy Monitoring and Targeting for Industrial Plants 253

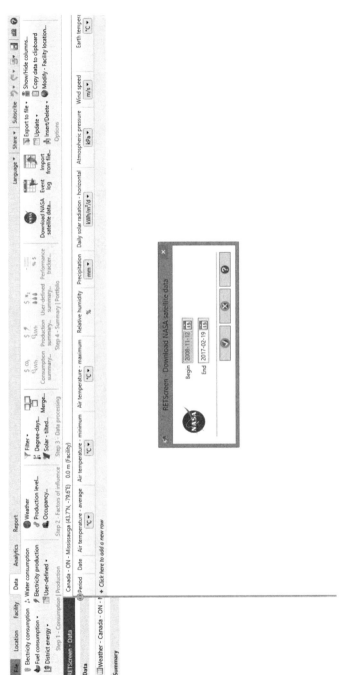

Figure 9-16. Selection of Time Frame for Weather

254 Industrial Energy Management Strategies

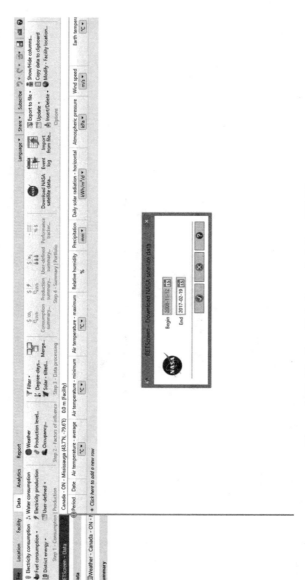

Figure 9-17. Weather Data Imported from NASA

Application of Energy Monitoring and Targeting for Industrial Plants 255

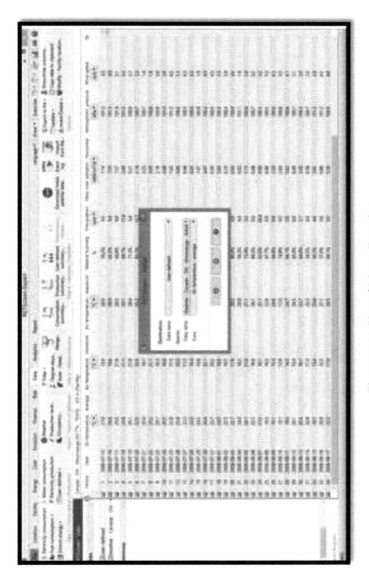

Figure 9-18. Use of Merge Function

256 Industrial Energy Management Strategies

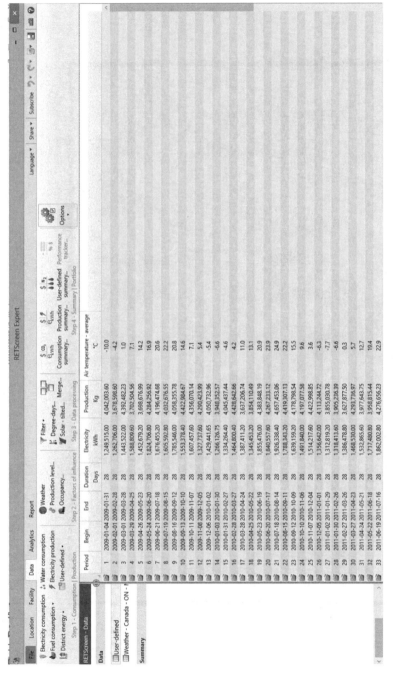

Figure 9-19. Data Preparation Complete

The next step is to click establish baseline and try regression analysis using production as an independent variable and electricity as the dependent variable. Reviewing the statistics of the regression model shows a poor R2, so adding another variable, which is average air temperature, can improve the R2, indicating a good acceptable model. Once a good correlation in the model is achieved, press the CUSUM tab in the analysis section.

Issues relating to CUSUM include the following:

1. *Selecting variables.* The variables selected have an effect on energy consumption, e.g., production, temperature, CDD, HDD, etc. Sometimes, more than one variable might be necessary to establish the baseline model.

2. *RETScreen Expert* has an event logger that can track all operational problems and corrective measures implemented.

3. *Selecting the baseline period.* For selecting the baseline model, it is always important to do an end-to-end CUSUM on the entire dataset and select the baseline period that has similar slope.

4. *The post-project period* might not show a proper regression as the energy system was not properly controlled. RETScreen Expert has a tab called "reference year," where it is possible to change the reference period from pre-project period to post-project period. In other words, if the project performance followed the post-project regression model, then the energy consumption in the baseline period would have been A. The difference in baseline energy consumption predicted from the reference period (post-project) and the actual finding would provide the energy savings.

5. Multimeasure issues, routine and non-routine adjustments. For whole facility analysis, it is important to eliminate the construction period. Select baseline and reporting period accordingly. A new addition of energy-consuming equipment during facility expansion might coincide with an energy project; while calculating energy savings, these adjustments for expansion need to be performed and are called non-routine adjustments.

258   Industrial Energy Management Strategies

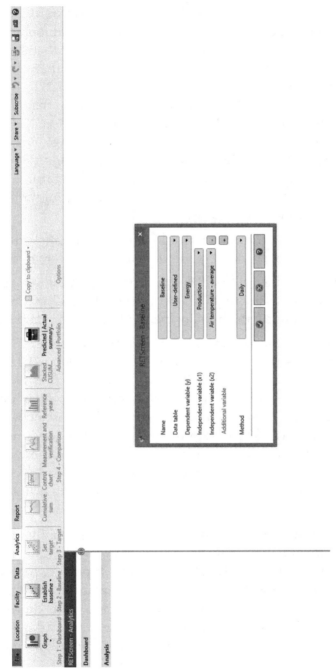

Figure 9-20. Selection of Variables for Regression

Application of Energy Monitoring and Targeting for Industrial Plants 259

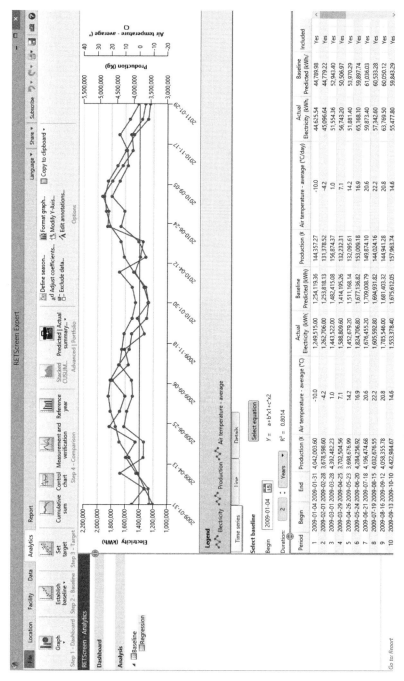

Figure 9-21. Two Variable Regression

260   Industrial Energy Management Strategies

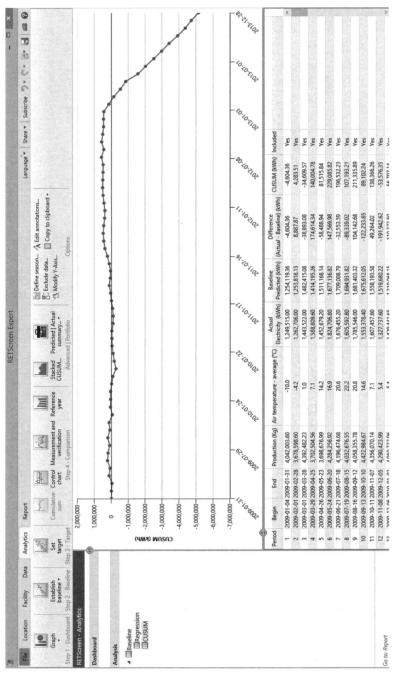

Figure 9-22. CUSUM Analysis

Application of Energy Monitoring and Targeting for Industrial Plants 261

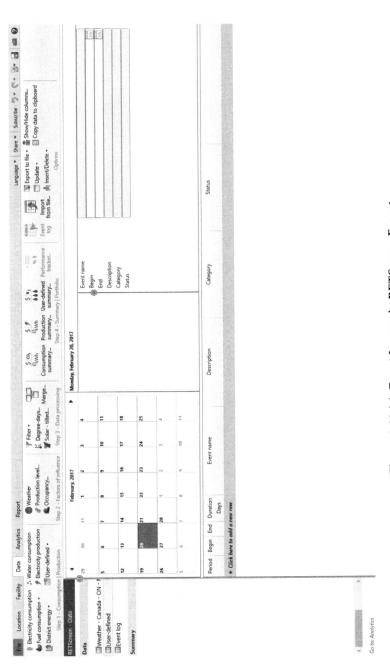

Figure 9-23 (a). Event Logger in RETScreen Expert

262   Industrial Energy Management Strategies

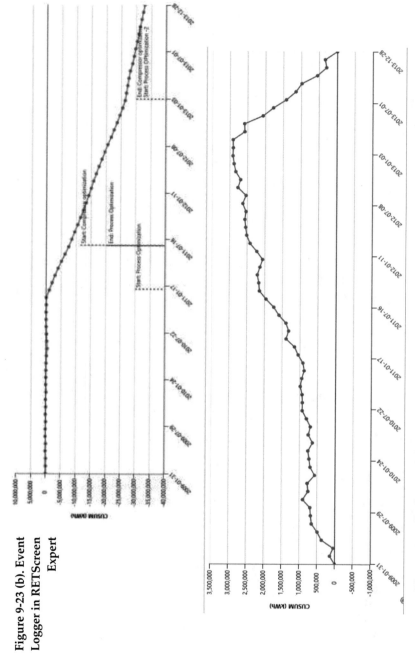

Figure 9-23 (b). Event Logger in RETScreen Expert

Figure 9-24. End-to-end CUSUM in RETScreen Expert (notice the points with similar slope for selecting baseline period).

Application of Energy Monitoring and Targeting for Industrial Plants 263

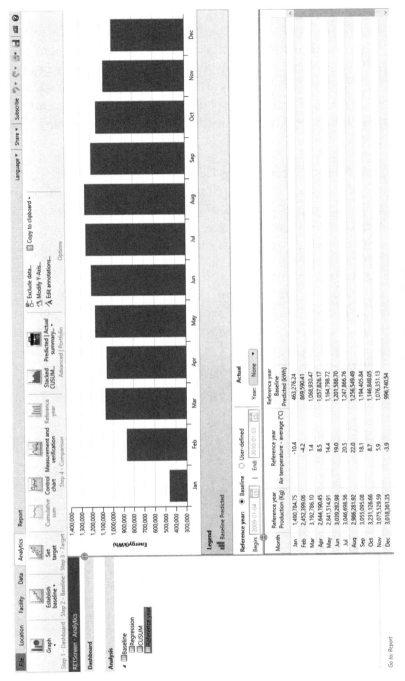

Figure 9-25. Reference Table

MONITORING AND TARGETING

Energy monitoring is an important energy management function. It allows for the detection of any change in the energy consumption pattern for energy equipment, an energy accounting center, or even the facility as a whole. Energy monitoring can help develop KPI (key performance indicator) and measure energy performance.

Setting up Energy Accounting Centers

Energy accounting centers can be equipment or complete-cost centers; for example, utilities would contain compressed air, steam consumption, etc. The energy accounting centers are based on significant energy use (SE), the way cost is treated in the plant.

Metering Requirements

The metering requirements can be based on energy consumption and the key drivers for energy consumption that would be used to develop a performance model to monitor consump-

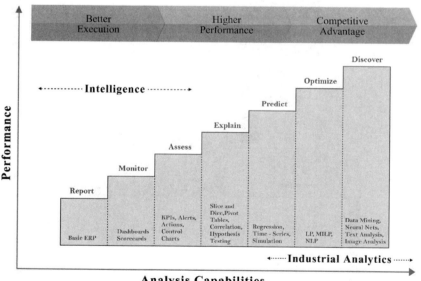

Figure 9-26. M&T Analysis

Application of Energy Monitoring and Targeting for Industrial Plants 265

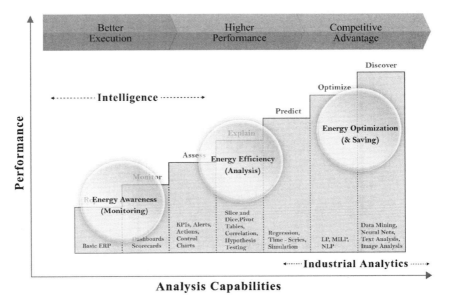

Figure 9-27. M&T Analysis

tion. In certain cases, several submeters and sensors are installed to measure energy performance of several subsystems. For example, in a chilled water plant, all components—namely, the chillers, the chilled water pumps, the condenser water pumps, cooling towers, and fans—are submeters along with individual chiller flow along with temperature differential across each of them. To develop kW per ton for chillers, the cooling tower also is part of the complete system.

Analysis

The data from the meters installed in an energy accounting center can be used to set performance targets. The performance targets can be in the form of KPI, such as kW per kg of product or system kW per ton for a chilled water plant. It can also be a percentage reduction compared to the energy baseline. The target can be set based on energy audits. In the following example a baseload reduction of 5% and production dependent energy reduction of 10% is targeted.

266   Industrial Energy Management Strategies

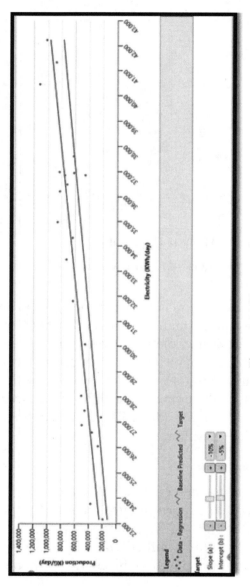

Figure 9-28. Target setting

The deviation from energy targets can be monitored at a specific time interval, and specific corrective action can be taken to address the deviation. **The following table** shows an energy target based on the performance model and the deviation from the target based on monthly energy and production information. **The −ve values shows** energy performance better than target while positive values show lower energy performance compared to the target.

CASE STUDIES ON DEVELOPING AN M&T SYSTEM FOR SMALL AND MID-SIZED FACILITIES

Small and mid-sized companies lack the metering infrastructure (except utility meters) to implement an M&T system. The case studies below show how an energy monitoring system was implemented based on the main utility meters.

Case A: A Stationery Product Manufacturing Company

This example involved a mid-sized manufacturing company. This company was losing market share, and cost reduction was critical. The company formed a team and conducted an energy assessment. Based on the energy assessment, it found that production equipment formed about 37% of energy consumption. However, lighting, compressors, and the trim system consumed more than 50% of the energy. The company decided to focus on the air compressors and lighting. Lighting measures, such as installation of motion sensors, delamping, compressed air leakage reduction, and installation of VFD on the compressors were identified. The company also decided to focus on measures that improved productivity.

An energy awareness campaign was launched at the company's health and safety meeting. Specific measures, such as shutting down conveyors during lunch hours, were discussed. The challenge was to quantify these initiatives. Monthly production data was collected with the corresponding energy consumption for the same period. The energy consumption was based on the main utility meter.

268   Industrial Energy Management Strategies

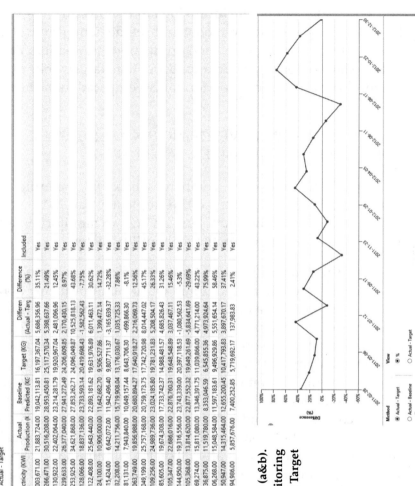

Figure 9-29 (a&b).
Monitoring
Against Target

A regression model was constructed. Fortunately, the regression between electricity consumption and production showed a strong correlation coefficient. Based on the regression model, a CUSUM chart was constructed. The CUSUM analysis showed that, as a result of the above initiatives, the company could reduce energy consumption by 218,048 kWh (7 months). The analysis helped demonstrate the results of the initiatives to management. M&T not only created confidence in the efforts but also helped the energy team secure funding for the capital projects.

Case B: Case of a Plastics Plant

A plastic manufacturing facility in Ontario, Canada, started implementing several energy conservation projects—namely, lighting upgrade and the installation of VFD compressors. They were interested in building an energy management program in the facility, and they formed an energy team. The team started looking at low cost and operational measures, such as reduction of compressed air leakages and pumping system optimization. Natural gas and electricity consumption from the utility meter was monitored monthly. A CUSUM analysis was carried out on a regular basis. This system of analysis could show the results of the team's energy/cost reduction efforts to senior management in terms of a reduction in the operating costs.

The energy team also was convinced of the value of regular monitoring and analysis. They decided to implement a low-cost EMIS (energy management information system). Major energy consumers in the facility—the boilers and compressors—were metered and production figures were entered manually in the system. A performance model was constructed and the energy consumption was reviewed daily.

270   Industrial Energy Management Strategies

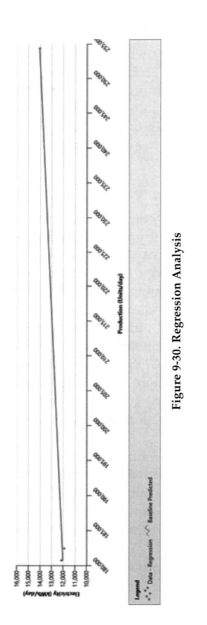

Figure 9-30. Regression Analysis

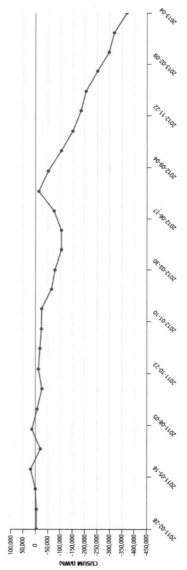

Figure 9-31. CUSUM Analysis for Low-cost Measures

Application of Energy Monitoring and Targeting for Industrial Plants

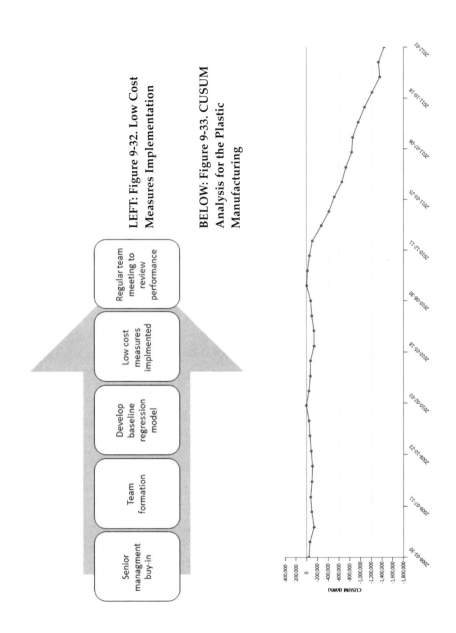

LEFT: Figure 9-32. Low Cost Measures Implementation

BELOW: Figure 9-33. CUSUM Analysis for the Plastic Manufacturing

Figure 9-34. Summary of M&T Implementation Stages

Case C: Case Study from a Steel Plant

A steel plant wanted to implement low-cost operational measures, and an energy team was formed. They identified a few issues, like equipment idling, compressed air leakages, and need for improvement in start-up and shutdown procedures. They also looked at scheduling production. The energy team wanted to quantify the operational savings and therefore collected monthly electricity data from the main utility meter and the corresponding tons of steel roll processed in the facility. A regression model was built, and the CUSUM analysis was conducted.

The steel plant carried out an audit and installed loggers on some of the idling equipment. Based on the analysis, some of the idling equipment was shut down. The CUSUM analysis was carried out on the electricity data obtained from the main interval meter, and it showed a reduction of about 436,000 kWh.

References

NRCan. RETScreen. http://www.nrcan.gc.ca/energy/software-tools/7465

Steve C. Schultz, C.E.M., C.L.E.P., Corporate Energy Manager, 3M, Beyond Metering, Increase Need for Energy Information System on Down Stream Production Equipment presented at the World Engineering Congress Conference, Washington, DC 2014.

Dollars to $ense Workshop: Energy Monitoring Workshop. Presentation. Natural Resources Canada.

Application of Energy Monitoring and Targeting for Industrial Plants   273

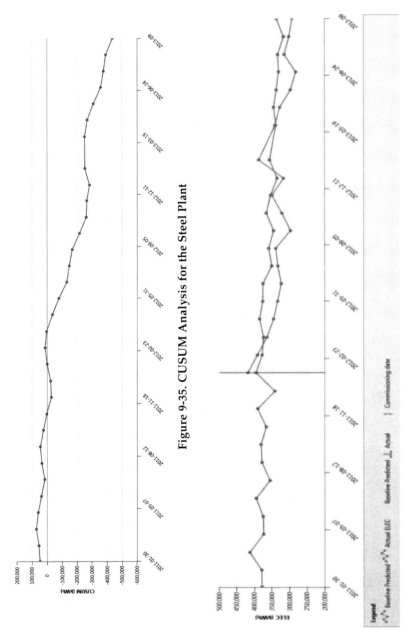

Figure 9-35. CUSUM Analysis for the Steel Plant

Figure 9-36. M&V for the Steel Plant Measures

274 Industrial Energy Management Strategies

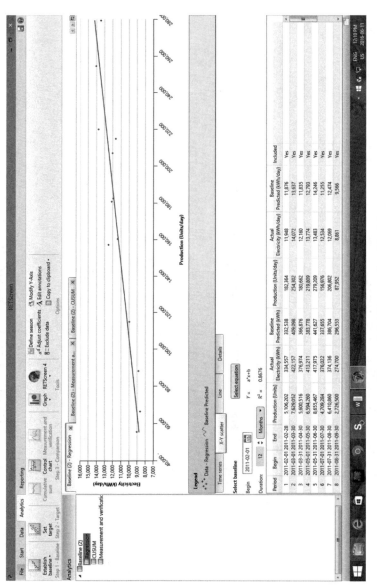

Figure 9-37. Regression Model

Chapter 10

Energy Management Innovation—Time-shared Energy Manager

INTRODUCTION

It has been widely accepted that any facility without an energy management program cannot get sustainable energy savings. While some of the large manufacturing companies have implemented such a program and benefitted, it is the small and mid-sized company that has several challenges in implementing such a program.

BARRIERS TO IMPLEMENTING AN ENERGY MANAGEMENT PROGRAM IN SME

Despite having significant energy efficiency potential, there are several key barriers that hinder the ability to achieve energy savings at small and mid-sized industrial facilities. These include:

- End users often lack the capital budget to self-finance energy efficiency investments.

- Efficiency projects at industrial facilities face the challenge of having to navigate an investment decision-making framework that places a heavy emphasis on optimizing manufacturing processes and ensuring continuous operation of plant assets.

- High ratio of transaction costs (i.e., conducting preliminary and detailed audits and establishing M&V protocols) to total project costs hinders the cost effectiveness of deals.
- Corporate capital budgeting processes place energy efficiency in direct competition with other core priorities, such as investments that expand production, increase throughput, and/or maintain overall plant reliability.
- Industrial firms have a short-term horizon for investments and typically require projects to have rapid payback periods.
- Lack of internal human resources to identify, execute, and verify energy conservation/efficiency projects.

THE ROLE OF THE ENERGY MANAGER

The time-shared energy manager would help in addressing specific barriers in the plant and demonstrate the benefits of an energy management program by achieving sustainable cost reduction. He/she would also start building capacity in the plant for sustaining such a program. The major activities of the energy manager would be developing energy benchmarks, selling energy management projects to senior management, carrying out energy performance analyses, demonstrating the value of energy and cost reduction to senior management, training and communication, help in formulating an energy team, identification of low-cost opportunities, working with other partners in the industry, and communicating the benefits to all stakeholders. Since the energy manager would be time-shared between various facilities, there can also be opportunities of aggregation between many smaller facilities. Some of the key roles of the time-shared energy manager are discussed in the following paragraphs.

Securing Funding

Corporate capital budgeting processes place energy efficiency in direct competition with other core priorities, such as

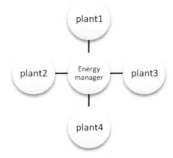

Figure 10-1. Timed-shared Energy Manager

investments that expand production, increase throughput, and/or maintain overall plant reliability. Therefore, securing internal funding for energy conservation projects may not be easy. The time-shared energy manager can address this barrier by preparing the business case and addressing the funding barrier by aggregating several smaller projects (from different industries) into larger projects to bring down the cost of some of the energy conservation projects. He/she can also tap into incentive programs that support energy conservation activities. There is also a possibility of partnering with an energy service company (ESCO) and implementing some of the measures through performance contracting. The ESCO can bring third-party financing for some of the energy conservation projects, which would allow quick approval by eliminating the need for internal funding. The ESCO may be able to monetize some of the carbon emissions credits from the energy conservation projects.

Building the Team and Implementing Best Practices

Building a cross-functional team is important to developing a shared understanding of the issues involved in managing energy of the facility and implementing best practices in their internal systems like procurement, operations and maintenance. The external energy manager would bring in the necessary energy management skills to work with the internal team to develop and implement the energy plan and review progress periodically.

The shared energy manager can also provide input for implementing process innovation. He/she can also advise the facility on some of the new opportunities, like ISO 50001 and other plant certification programs. Regular roundtables can be organized by the time-shared energy managers where representatives from all clients can participate and network to share experiences and identify new avenues of cooperation.

CASE STUDY OF THE TIME-SHARED ENERGY MANAGER

The SaveONEnergy program in Ontario is designed to address barriers and encourage electricity consumers to implement energy conservation opportunities in the residential, commercial, institutional, and industrial sectors. The program ranges from providing financial incentives for various equipment upgrades to access to a shared energy manager. This section is primarily limited to the discussion of the role of the shared (roving) energy manager (REM) and how it helps customers in addressing barriers.

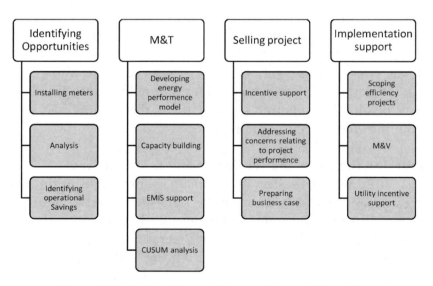

Figure 10-2. Roving Energy Manager Support Example

Identification of Good Energy Conservation Opportunities

There are often several low- and minimal-cost energy conservation projects than can be implemented to reduce energy in an industrial facility. In many cases, the plant is aware of some of the opportunities, but in the absence of quantification of the energy savings and investment required, the relative importance of measures are overlooked. The REM helps customers to prepare the business case for such measures. The following are examples of some of the low-cost/lower payback energy conservation opportunities.

Assistance towards Identifying Building Automation System (BAS) Retro-commissioning

Operational improvement measures like BAS recommissioning often lead to better comfort and associated energy savings. The REM works with BAS contractors and operators to identify and implement some of these low-cost measures.

The REM reviewed the building automation system (BAS) in a large manufacturing office. The following energy reduction opportunities were identified:

1. Addition of chilled and condenser water resets.
2. Updating exhaust fan operating schedules.
3. Trending of speed for VFD drive installed in the chilled water pumps.
4. Improvement of outside air controls for the rooftop units.

It was also recommended that the permission to change setpoints in the system be limited to a few individuals who operated the system.

Demonstrating the Value of Energy Management by Monitoring Facility-wide Energy Consumption

Energy monitoring can help in understanding how energy is used in the facility, and the relationship with variables like production and weather. This relationship can form the basis of measuring progress by carrying out a CUSUM (Cumulative Sum) analysis.

As an example, a stationary product manufacturing plant was losing market share and therefore cost reduction was very critical. The company enrolled the REM to provide assistance in reducing energy costs. The REM worked with the team and identified opportunities. The company decided to focus on operational measures. It also implemented some low-cost measures like installation of motion sensors and BAS recommissioning. An energy awareness campaign was launched at the company's health and safety meeting. Very specific measures, like shutting down conveyors during lunch hours, were discussed. It was important to measure progress and demonstrate results of the initiatives to get the senior management buy-in for some of the larger projects.

A linear regression between production and energy was developed, and we found a high regression coefficient ($R2=$). This was used as a baseline model ($y=0.0206X+7658$) which is in the form $Y=MX+C$. The regression model was then used as a basis for evaluating monthly energy performance for the facility. CUSUM analysis was carried out and an electricity savings of 218,048 kWh was identified.

Support in Implementing Small Projects

The REM also provides assistance (budgeting, planning, developing specifications, etc.) in moving the energy project towards implementation.

A plastics company enrolled an REM to provide assistance in getting incentive and help to implement a pumping retrofit project. The plant had several water-cooled molding machines. The cooling water system had tanks, circulating pumps, heat exchangers, and a cooling tower. The molding machines were cooled by the recirculated cooling water system. Water heated from the molding machines was stored in a tank, which was then passed through a heat exchanger by the heat exchanger pump. After passing through the heat exchanger, cold water was stored in the process tank. The water from the process tank was pumped back to the

Energy Management Innovation—Time-shared Energy Manager 281

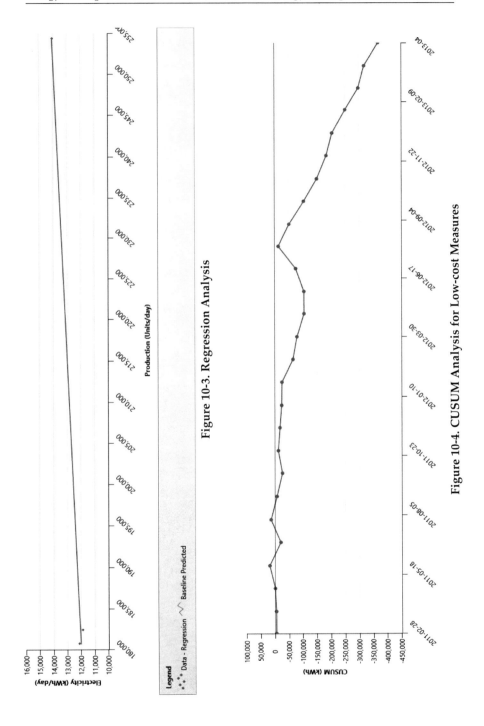

Figure 10-3. Regression Analysis

Figure 10-4. CUSUM Analysis for Low-cost Measures

282   Industrial Energy Management Strategies

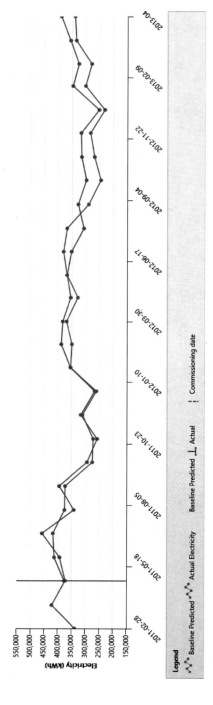

Figure 10-5. M&V for Low-cost Measures

molding machines by the process pumps. Three out of the four process pumps were operated. The hot water in the secondary side of the heat exchanger was cooled in the cooling tower.

The REM reviewed the system and recommended that a pumping study be carried out. The study involved measurement of flow, pressure, and input power at all the pumps in the system. It was found that the heat exchanger pumps were operating at very low efficiency. Additionally, the flow and power measurements at all the process pumps showed that one out of the three pumps was delivering very low flow compared to the input power, and therefore it was decided to operate two out of the three process pumps. The REM worked with other service providers and contractors to determine some of the project specifics, like project savings and implementation cost. The team also determined some of the piping modifications that were required to implement the project.

Figure 10-6. Energy Project Support

Similarly, a large laundry company enrolled an REM to implement some energy conservation projects. The REM identified several projects. Two projects were prioritized for implementation. One was the compressed air retrofit, and the other was blower optimization. The compressed air retrofit project involved installation of a VFD air compressor. The savings for the project was calculated to be 50 kW demand and 230,000 kWh with a simple payback of 0.9/year with incentive. The REM installed amperage loggers to determine the compressor load profile. DOE–Air Master was used to calculate the base case air demand and energy consumption. The energy efficient case consumption was based on a compressed air performance curve. The REM collected all the necessary information such as quotes and performance curves, and reviewed energy savings with the energy team.

Figure 10-7. New Compressor at the Laundry

Support in Selling the Energy Conservation Projects

Several good (low payback) energy conservation projects do not move forward due to concerns like operational, reliability, safety and maintenance. The REM assists consumers in addressing some of these concerns.

In a plastics company, a compressed air project was already identified by the vendor. The project involved re-piping a compressed air distribution network to lower the operating pressure. In the process, the proposed plan was to shut down one 75 HP compressor. The plant had concerns regarding the system pressures. The REM reviewed the project, and a pilot test was conducted jointly with the compressed air vendor, the plant maintenance engineer, and the REM. The pilot test involved simulation of the lower pressure piping system by connecting different networks by opening an interconnecting valve. The compressors were run at a lower pressure, and amp loggers were installed in all compressors. The system pressure was also measured at various points, and it was revealed that the plant could indeed lower

Energy Management Innovation—Time-shared Energy Manager 285

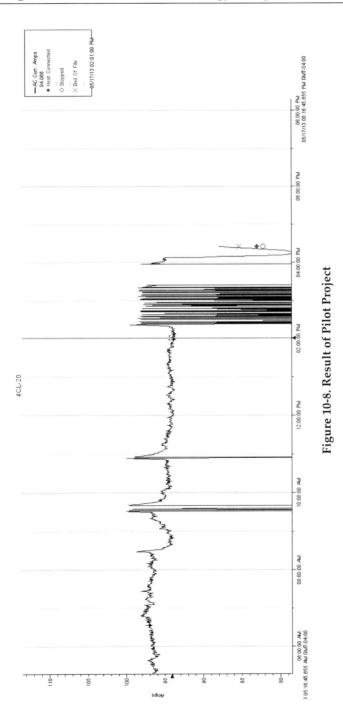

Figure 10-8. Result of Pilot Project

the pressure and one of the compressors could be shut down. The plant had a budget limitation, and the project scope was modified accordingly.

Capacity Building

A plastics manufacturing facility in Ontario, Canada, started implementing several energy conservation projects, namely lighting upgrades and the installation of VFD compressors. They were interested in building an energy management program in the facility. They formed an energy team with the REM. The team started looking at low-cost and operational measures, such as reduction of compressed air leakages and pumping system optimization. Natural gas and electricity consumption from the utility meter was monitored on a monthly basis.

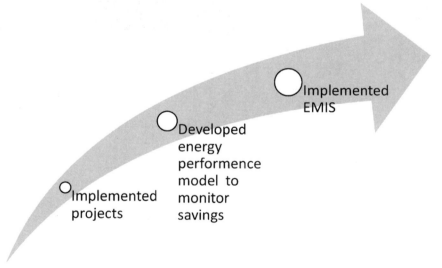

Figure 10-9. Capacity Building for EMIS Implementation

CUSUM analysis was carried out on a regular basis. This system of analysis could show the results of the team's energy/cost reduction effort to the senior management in terms of reducing the operating cost.

Energy Management Innovation—Time-shared Energy Manager 287

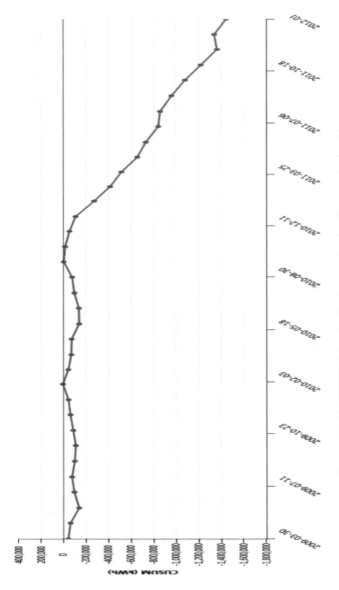

Figure 10-10. CUSUM Analysis Demonstrating Energy Reduction

The energy team also was convinced of the value of regular monitoring and analysis. They decided to implement a low-cost EMIS (energy management information system). Major energy consumers in the facility, namely the boilers and compressors, were metered. Production figures were entered manually in the system. A performance model was constructed, and the energy consumption was reviewed on a daily basis.

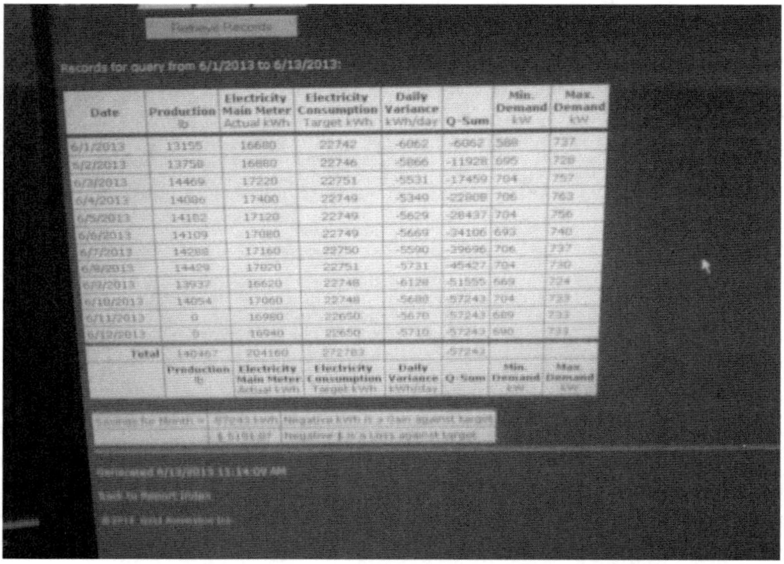

Figure 10-11. Energy Management Information System

References

Bhattacharjee, Kaushik (2013). Effective use of utility incentives to develop an industrial energy management program, presented and published at World Engineering Congress Conference, Washington, DC.

Bhattacharjee, Kaushik (2009). Industrial energy management-time shared energy manager program, presented and published at World Engineering Congress Conference, Washington, DC.

https://www.saveonenergy.ca/Business/Program-Overviews/Energy-Managers.aspx.

Index

A

action plans 222
actual energy consumption 233
AEE 169
affinity laws 32
afterburners 40
air compressors 267
air distribution network 284
air flow meters 5
air leakage 70, 193, 272, 286
Air Master 68
ammeter 3
amp measurements 3
analyze energy use 213
Andon 79
annual energy use patterns 2
application engineering issues 45
assembly line operations 95
Association of Energy Engineers 227
availability 81

B

barriers 210, 226
BAS recommissioning 173
baseline equation 231
baseline period 176
batch operation 95
benchmark 199
best efficiency point 171
best practices 139, 277
bin analysis 27, 33
bin data 30
blowdown heat recovery 36
blower optimization 283

boilers 149, 288
Btu/part 89
Btu/ton 88
budgeting 155, 208, 280
building automation system 279
building operators 199

C

capacity building 286
capital budget 155
cascade operation 201
cash inflow negative 158
catalytic recuperative oxidizer 42
CDD (cooling degree days) 2
chilled water optimization 199
chilled-water system 44
    analysis tools (CWSAT) 47
chiller bank optimization 46
chillers 226
    sequencing 45, 231
coffee roasters 44
cold storage 60
combustion analyzers 6
commissioning agent 202
commissioning projects 208
common operational measures 141
compressed air 284
    retrofit 283
    system 36, 68, 146, 198
compressed system 231
compressors 3, 178, 193, 201, 226
    air 227
    cooling water 53
    inlet 147

condenser water resets 279
condenser water temperature 196
condensing economizer 39
condensing pressure 56
consultants 210
continuous improvement 84, 206
continuous manufacturing 95
contractors 283
conveyors 145, 173
cooling tower 52, 280
cooling water system 200
correct matrix 157
cross-functional team 277
current transformer 5
CUSUM 199, 229, 280, 286
    analysis 151, 173, 202, 227, 230, 234, 272
    concept 232

D
data logging 9
data preparation 237, 248
defects 95
demand measures 193
detailed documentation 199
direct contact hot water heater 37
direct fired thermal oxidizer 41
discharge air temperature setpoint 29
distributed control system (DCS) 177
driver 2
dry bulb temperature 2
dust collection system 32

E
ECAM 16

efficient refrigeration system 58
embedded energy manager 210
employee engagement 209
energy accounting centers 264
energy awareness 267
    campaign 280
energy baseline 219
energy benchmarks 2, 276
energy champion 155, 226
energy conservation projects 277
energy continuum 206
energy efficiency of pumps 64
energy efficiency projects 226
energy-efficient chillers 46
energy improvement 217
energy intensity 88
energy management best practices 205
energy management continuum 205
energy management information system 229, 288
energy management initiatives 227
energy management leaders 206
energy management program 178, 212, 275
energy management review contour 224
energy managers 278
energy measurements 3
energy monitoring 279
    system 208
energy objectives 222
energy performance indicators 219, 220
energy policy 222
energy/process flow diagrams 13

Index

energy project support 283
energy team 140, 272
Energy Traffic Light Program 141
energy use breakdown 10
equipment effectiveness, overall 79, 80
equipment idling 142, 272
ESCO 277
evaporative condensers 60
evaporator coil 58
evaporator fans 58
EVO 169
Excel 249
exhaust system 30

F
faults 139
field trials 196
financial analysis 201
financial incentives 210, 278
financial matrixes 157
floating head pressure control 56, 201
free cooling 193
free money 158
fully productive time 82
funding barrier 277

G
gas consumption 149
greenhouse gas emissions 211

H
health and safety 267
heat exchangers 280
heating energy consumption 26
heat recovery 36, 60
    from process reactors 38

I
internal rate of return 157
implementation cost 283
industrial cooling water system 50
industrial refrigeration 55
industrial ventilation systems 24
industry associations 210
installation of variable flow chilled water pumps 48
insulation 29, 30
interconnecting valve 284
internal rate of return 157
International Performance Monitoring and Verification Protocol (IPMVP) 172
interval data analysis 15
interval meter 272
IPMVP options 170
ISO 50001 169, 211, 278

K
Kaizen 79, 84, 98
key performance indicator 264
KPI 229
kWh/linear yard 88
kWh/pound 89

L
lean characteristics 79
lean manufacturing 77
LED lamps 60
lighting 267
    controls 176
    retrofit 64, 173
light meter 8
linear regression 173, 280
logged amps 143
loggers 172, 196, 284

logging 3, 13
low cost energy loggers 8

M
machine interlocks 145
maintenance issues 139
make-up air 25
manual valves 199
measurement 213
    method 13
measurement and verification
    169, 200, 236
    plan 176
metal halide 60
metal processing plant 52
metering 196
    plan 200
    requirements 264
milling machines 52
molding machines 201, 280
monitoring and targeting 264
M&T analysis 265
M&T system 267
Muda 79
M&V plan 176

N
nameplate method 12
NASA 234
net present value 157
new chiller 158
non-critical loads 145
non-productive energy 89
normalization 172

O
occupancy logger 142
occupancy sensors 142
operational abnormalities 208
operational measures 139
operational savings 139
operation protocol 202
operator training 199
optimize motor size 149
optimizing cooling water 51
optimizing ventilation systems
    33
organizational 206
organizational structure 224

P
packaging machines 145
partnership 209
payback 284
PC/PLC 113
performance 81
    matrix 208
    model 233, 288
    targets 265
pilot 284
    projects 196
piping system 200
pivot tables 15
planning 280
plant operating time 82
plastic extrusion plant 14, 25
plastics manufacturing 198
portable water flow meters 6
post-project operation 198
post-project period 257
power analyzers 3
power measurement 52
predicted energy consumption
    233
preheat combustion air 38
premium efficiency motors 218
pre-scoping 192
    stage 196

Index

pressure decay test 74
prioritize 217
procedural control 218
process control 196
process heat recovery system 40
process pumps 283
product innovators 209
production process 196
productive energy 89
project-based approach 224
project commissioning 202
project costs 156
project delays 201
project development stages 192
project pitfalls 191
project prioritization 192
    spreadsheet 192
projects 156
project scope 202
PRV 149
pump 197, 280
    efficiency 66

Q
quality 82

R
reciprocating compressors 45, 193
recommissioning 199
reduce idling 143
redundancy 197
regenerative thermal oxidizer 43
regression analysis 236
regression models 2
regulatory compliance 228
repair 158
reporting period 176
reset controls 47

retro-commissioning 279
RETScreen Expert 2, 18, 73, 151, 227, 237, 247, 257
rewarding employees 140
roaster 40
rooftop units 279

S
SCADA 113
SCADAware 113
scatterplot 231
scoping 196
scrap 91
    rates 95
securing funding 276
selecting variables 257
selling 156
senior management commitment 206
sequencer 68
short-term horizon 155
significant energy users 215
simple payback 227
small and mid-sized companies 226
spot measurements 9, 172
steam boiler 37
steam pressure optimization 149
steam traps 140
strategic approach 191
submetering 218
    system 199
submeters 140, 198, 265
suction air 147
superior energy performance 211
supply measures 193
sustainable energy management 191
sustainable energy program 206

sustainable energy reduction 224
system optimizations 198
systems approach 158, 193

T
targets 222
technological 206
temperature bin hours 26
thermal recuperative oxidizer 42
third-party audits 212
third-party financing 277
time-shared energy manager 275
Toyota Production System 87
training operators 140
transaction costs 155

U
ultrasonic leak detectors 8

utility bills 2
    analysis 2
utility incentive 158
    programs 209, 210

V
vacuum pumps 142, 145
variable-flow xhilled and condenser water pumping 45
variable frequency drive 218
VFD compressors 286

W
water-side economizer 46
WattNode Pulse energy meter 218
weak management commitment 226